OCEAN GREENS

SAMPHIRE

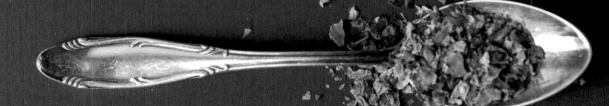

PERSIL DE LA MER

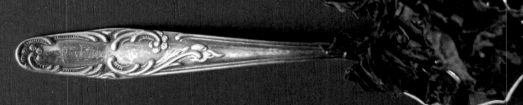

WAKAME

OCEAN GREENS

Explore the World of Edible
Seaweed and Sea Vegetables

A Way of Eating for Your Health
and the Planet's

WITH 50 VEGAN RECIPES

Lisette Kreischer
Marcel Schuttelaar
and others

THE EXPERIMENT
NEW YORK

The contents of this book have been carefully considered and reviewed by the authors and publishers, yet we cannot guarantee the food safety and quality of algae and algal products. The author and publisher specifically disclaim all responsibility for any liability, loss, or risk—personal or otherwise—that is incurred as a consequence, directly or indirectly, of the use and application of any of the contents of this book. The authors recommend a daily dose of no more than 10 grams of (dried) algae per day, comprised of a variety of species. The authors strongly advise readers to contact local producers and the relevant food authority for the safety of seaweed and algae products.

Library of Congress Cataloging-in-Publication Data

Names: Kreischer, Lisette, 1981- author, editor. | Schuttelaar, Marcel author, editor.
Title: Ocean greens : explore the world of edible seaweed and sea vegetables : a way of eating for your health and the planet›s / [compiled and written] by Lisette Kreischer & Marcel Schuttelaar, et al.
Other titles: Groente uit zee. English
Description: New York : The Experiment, [2016]
Identifiers: LCCN 2016030994 (print) | LCCN 2016031244 (ebook) | ISBN 9781615193523 (cloth) | ISBN 97816156193530 (Ebook)
Subjects: LCSH: Cooking (Marine algae) | Marine algae as food. | Marine algae. | LCGFT: Cookbooks.
Classification: LCC TX402 .G7613 2016 (print) | LCC TX402 (ebook) | DDC 641.6--dc23
LC record available at https://lccn.loc.gov/2016030994

ISBN 978-1-61519-352-3
Ebook ISBN 978-1-61519-353-0

Research and text by Marcel Schuttelaar, Koen van Swam, and Susanne Hagen
Recipes and photography by Lisette Kreischer
Recipe styling by Roos Rutjes
Cover and text design by Sarah Schneider

Additional Photography: Stichting Noordzeeboerderij (page x), Geert Gordijn (page 180), Hortimare (page v), Dos Winkel (pages xii, 1), Wild Wier (pages 5, 31), Guy Buyle (page 32), Maurits Bos (photo of Lisette Kreischer, page 180), NASA Earth Observatory image by Jesse Allen (page 7), Marieke Eyskoot (page 13)

Recipes copyright as follows: First published by Succesboeken in *Non*fish*a*licious* © 2011 Lisette Kreischer and Sea First Foundation: Non*Fish Sauce (page 59), Eggplant Caviar (page 80), The Fabulous Fishless Soup (page 96), Cut the Crab Salad (page 104). First published by Kosmos in *Plant Power* © 2014 Lisette Kreischer as Ocean Breeze (page 95).

Manufactured in China
Distributed by Workman Publishing Company, Inc.
Distributed simultaneously in Canada by Thomas Allen & Son Ltd.

First printing October 2016
10 9 8 7 6 5 4 3 2 1

DEVIL'S APRON
(*SACCHARINA LATISSIMA*)

Contents

WAKAME   SEA LETTUCE   DULSE

Author's Note

I LOVE THE NORTH SEA; the shoreline runs right next to where I live. Leaving the harbor in my sailboat, I've crossed its waves numerous times. On the map it may seem like a modest body of water, but once at sea you're struck by its vastness, the emptiness of the place. Besides loving the sea I also have a fascination with vegetables: the enormous variety of types, with their many different colors and flavors. We can all afford to eat more greens. I first researched the consumption of seaweed about forty years ago as a student of human nutrition. Seaweed, I was convinced, would solve the world's food problem. As it turns out, I was a little ahead of the curve back then.

Several years ago pieces of the puzzle started to come together. I was helping Willem Brandenburg, a preeminent scholar on seaweed, organize a seminar on seaweed cultivation. It was Mr. Brandenburg who pioneered this cultivation in the Dutch province of Zeeland. After all these years, I became reacquainted with the subject and that's when something began to grow. . . .

Soon after, I was asked to help grow seaweed in the North Sea. Would my company be interested in investing energy, time, and money in the founding of a North Sea farm? Despite ample research and lots of talk about seaweed in the Netherlands, no actual cultivation was taking place in the North Sea. It felt right, and we got started. A group of young and enthusiastic people went at it with tremendous energy, often volunteering much of their spare time. The North Sea Farm Foundation was revived, and the first experiment with seaweed farming in the North Sea became a success.

During my first trial at sea I got to know Lisette Kreischer and Roos Rutjes: two women who had chosen to follow a plant-based diet and to become seaweed ambassadors. They wanted to put together a seaweed cookbook and were looking for a partner. In order to put seaweed fully in the spotlight we decided to include only purely plant-based recipes. This is what makes it such a special book. The title was obvious: *Ocean Greens*. It's a book in which we share our knowledge and skills so that anyone can enjoy eating seaweed as much as we do.

This book deals with algae and plants that thrive in saltwater—and there's plenty of that all over the world. It features some of the world's seaweed pioneers: Bren Smith, an ocean farmer and CEO of GreenWave (page 70); Shep and Seraphina Erhart, owners of Maine Coast Sea Vegetables (page 86); Kathy Ann Miller, curator of algae at UC Berkeley (page 98); Mark Kulsdom, cofounder of The Dutch Weed Burger (page 114); and Prannie Rhatigan, a seaweed chef (page 142). It also covers the different species and where they grow, their health benefits, how great these plants taste, and how to turn them into the most wonderful dishes.

Seaweed is a term for a variety of remarkable organisms with many illustrious names like dulse, kombu, wakame, Irish moss, devil's apron, egg wrack, bladderwrack, sea oak, sea lettuce, and sea spaghetti. There are thousands of green, red, and brown varieties, each of them different and many containing as-yet unknown organic compounds. They're so full of bioactive nutrients that you should consume them with some moderation, which has been taken into account in this book. With their multitude of extraordinary fibers and proteins, wonderful algae-derived oils, and many minerals and spores, they are the perfect supplement to our daily diet. They have long been a staple of Asian cuisine for centuries and have long been a part of cultural rituals and medicinal practices—so look for them in Asian markets. In recent years, their characteristic umami flavor has become increasingly popular in the West, and now you, too, can cook with it.

Ocean Greens also covers vegetables that grow on the border between salt water and freshwater in brackish coastal areas that get inundated once in a while. From marsh samphire and common ice plant to sea beet, there's still so much to discover.

This book arrives at a special moment in time. For centuries the sea has been perceived as a threat: an enemy that had to be kept at bay. Salinization was the nightmare of every farmer. Now we find ourselves at a turning point: Salt is on the rise, and more and more, the blue oceans are becoming a green source of both sustainable fuel and tasty food for our future.

Marcel Schuttelaar
Chairman, North Sea Farm Foundation
Founder, Schuttelaar & Partners, consultancy for a sustainable and healthy world

> # WE'RE AT A TURNING POINT; SALT HAS WIND IN ITS SAILS.

OCEAN GREENS

SOUTH AUSRALIA,
KANGAROO ISLAND

SEAWEED: WHAT IS IT?

SEAWEED AND ALGAE can be found all over the world. Like bacteria, algae were among the first forms of life on earth. Algae survive under the most severe circumstances, and many researchers believe we owe our existence to the development of algae billions of years ago. That is because algae practice photosynthesis (like trees and plants do), releasing oxygen into the atmosphere and enabling life as we know it. There are thousands of species of seaweed that can differ from one another like day and night. Some kinds grow very slowly and remain small; others create entire forests, so-called "kelp forests," in the water. Most types of seaweed are edible, though their flavors vary widely. For example, sea lettuce (*Ulva lactuca*) is a tender green seaweed that is very accessible, while other kinds are much tougher and not as easy to eat. In Asia, seaweed has been a staple for years and is considered a delicacy. In the United States and Europe, the popularity of seaweed is steadily on the rise. Just a few years ago, not many people were looking for it in their local grocery stores. Nowadays, it's everywhere, being sold in most grocery stores in some variety.

What is seaweed?

Algology, or phycology, is the scientific study of algae. *Phyco* (*phukos* in Greek) means "seaweed" and *logy* (*logos*) is defined as "knowledge" or "science." *Seaweed* is the umbrella term for the entire group of macroalgae that live in salt water, brackish water, or freshwater. Algae can be divided into two groups: microalgae (unicellular organisms, like chlorella) and macroalgae (multicellular organisms, like seaweed).

We already know about tens of thousands of species of algae, and new species are being discovered constantly. From an evolutionary perspective, algae are very old. It's estimated that multicellular seaweeds came into existence about 1.2 billion years ago. Unicellular algae have been around for even longer, about 2 billion years. Fossils of algae that date back 600 million years have been found. Seaweed species are usually categorized among the cryptogams (plants that do not produce flowers or seeds), like ferns. They resemble ferns in terms of their structure and reproduction, and are more elementary in that respect than flowers, trees, or agricultural crops.

In terms of their build, seaweeds do resemble plants, with a "root," a stalk, and leaves. There are a few big differences, though. The root of seaweed, called the holdfast, isn't used to absorb water and nutrition from the soil as with land plants, but rather to cling to rocks, shells, and other weeds. Not all seaweeds cling to a fixed object—some algae float in the water. Connected to the holdfast is a stalk that supports the leaves of the weed. Because seaweeds have a more elementary structure than terrestrial plants, which are at a more advanced degree of evolution, they are able to grow faster and produce more biomass than their terrestrial neighbors.

There are three groups: green algae, brown algae, and red algae. The color is determined by the composition and amount of pigment present in the algae, which differs by group. Pigments are chemical compounds that reflect light and are perceived as color. Photosynthetic pigment captures light for photosynthesis, and likewise differs by group. Algae that live in deeper waters require more pigment to absorb sunlight than algae growing just below the surface. *Laminaria abyssalis,* a relative of the oarweed, has been found at a depth of 360 feet (110 m). Red algae can live in the deepest water, as they are capable of absorbing energy from the sun at very deep levels. Some algae are able to live as deep as 820 feet (250 m) below sea level if the water is exceptionally clear; only 0.0005 percent of sunlight comes through at this depth. Aside from the above-mentioned groups of algae—the red, brown, and green algae—there is a fourth group: blue algae, also called blue-green algae. Blue algae aren't actually algae but bacteria, specifically cyanobacteria. Most of these algae are poisonous. A tasty exception is spirulina.

WAVES

When the leaves of seaweed grow very large, they are hindered more by waves, in the way that wind affects a large umbrella. Increased movement in the water creates more pressure on the holdfasts of the seaweed. The leaves might tear, though this doesn't pose any problem to the algae, which will continue to grow, undisturbed, and because of their torn leaves, the waves will bother them less.

Where can you find seaweed?

Seaweed grows in every sea or ocean, from the warmer waters around the equator to the ice-cold areas around the poles. Algae need water, sufficient sunlight, nutrients, and sometimes an "anchor" to attach themselves to. Seaweed uses light energy from the sun (photons) to grow, and this happens with help from photosynthetic pigments. During photosynthesis, water and carbon dioxide (CO_2) are converted into sugars with light energy derived from the sun. Oxygen (O_2) is an important by-product of this process.

Because seaweeds need light, they usually grow near the water surface. Most seaweeds live at a depth of up to 16 feet (5 m) underwater. There is hardly sufficient light below that level. The majority of seaweed species grow near coastal areas and rocky seashores, but

A large part of all photosynthetic processes on earth is carried out by algae. Forests and plants aren't the only organisms providing oxygen, as commonly thought. Seas and oceans also play a large role. The percentage contributed by algae and seaweeds to the oxygen supply differs in the scientific literature but is nonetheless staggering. Some scientists believe that half of our oxygen is derived from algae; others say it's 90 percent.

some species float freely in the ocean. There are species that always need to stay submerged because they'll die once they dehydrate. Other species can survive without water for a while, during low tide, for example. Seaweeds can form sprawling forests in the water with plants that reach several feet long. These kelp forests form a world on their own: Fish and other sea creatures live in these underwater habitats, using the algae as protection from open water.

Life cycle

The life cycle and reproduction of seaweed species differ from those of regular plants. Different species of seaweed are distinguished by their cycles, though we don't know the origin, growth, or development of many kinds. With kelp, a group of brown algae, the cycle starts when mature weeds develop spores. The spores develop into gametes, which are either male or female. When these gametes are fertile, the males produce sperm cells, the females produce egg cells, and they fuse into a "sporophyte" or "baby alga." The cycle comes full circle when young algae grow into mature algae and start producing spores for future generations.

Then there's vegetative reproduction, which occurs in some algae. It happens because of vegetative growth: when a piece of algae—a leaf, for example—tears off and fosters the growth of an entire new seaweed. This is comparable to replanting cuttings of plants that grow on land. In these cases, the seaweed grows thanks to nutrients in the water, crucially the elements phosphorus and nitrogen.

Green algae (Chlorophyta spp.)

Green algae derive their name from their color. The Chlorophyta can be found as multicellular organisms or unicellular ones—of which there are 1,500 to 2,000 species! Green algae live in both salt water and freshwater, and the unicellular algae particularly prosper on the coast, just above the flood line. These algae prefer nutrient-rich waters: High concentrations of phosphate and nitrate (due to pollution) are good for them. Green algae contain the same photosynthetic pigments as land plants, chlorophyll a and chlorophyll b. A well-known green algae is the *Ulva*, or sea lettuce, which lives near the east and west coasts of the United States, among other waters.

WAKAME

Kathleen Drew-Baker deciphered the life cycle of the red algae *Porphyra*—the famous nori seaweed in sushi—in the 1940s, enabling people to cultivate these red algae on a commercial scale. A couple of years before her discovery, Japan was hit by a number of typhoons, which, in combination with the contamination of water at the coast, brought a complete halt to the growth of nori. At the time nobody was familiar with the life cycle of nori, and it was therefore impossible to cultivate new nori and reintroduce it to the area. Kathleen Drew-Baker's discovery in 1949 and her subsequent publication in *Nature* magazine made it possible to start the cultivation of this seaweed after the typhoons.

Brown algae (Phaeophyta spp.)

Brown algae are almost exclusively saltwater species and are rarely found in freshwater. Most kinds live in the cold waters of the Northern Hemisphere. One of the largest algae of any type is a brown alga: the giant kelp (*Macrocystis pyrifera*), which can grow up to 200 feet (60 m) long and is found near southern Alaska. Another brown alga is the sargassum, which creates unique ecosystems in the tropical waters of the Sargasso Sea. In the North Sea, one finds mostly oarweed (*Laminaria digitata*) and devil's apron (*Saccharina latissima*). Worldwide, there are an estimated 1,500 to 2,000 species of brown algae.

Red algae (Rhodophyta spp.)

Red algae vary greatly in color: from soft pink to dark red, purple, and even a black-red. The color is determined by the amount of red pigment in the seaweed. These algae do have green pigment (chlorophyll), but they're dominated by the red pigments (phycocyanin and phycoerythrin).

About 6,500 species of red algae are known. Most red algae can be found in salt water, but a few species are known to exist in freshwater. Because they absorb the blue light from the sunlight spectrum better than green or brown algae do, they're better suited to deeper waters. One can also find red algae just below the surface. In general, red algae grow more slowly than species in the other two groups. They mostly live in warmer waters and can be found in abundance in Asia.

Red algae are most used for consumption. The seaweed in sushi, nori, is a red algae. It's often thought that nori is green algae because of its color in food. However, nori turns that greenish-brown only as a result of drying and roasting during processing.

SEAWEED FIELDS,
SOUTH KOREA

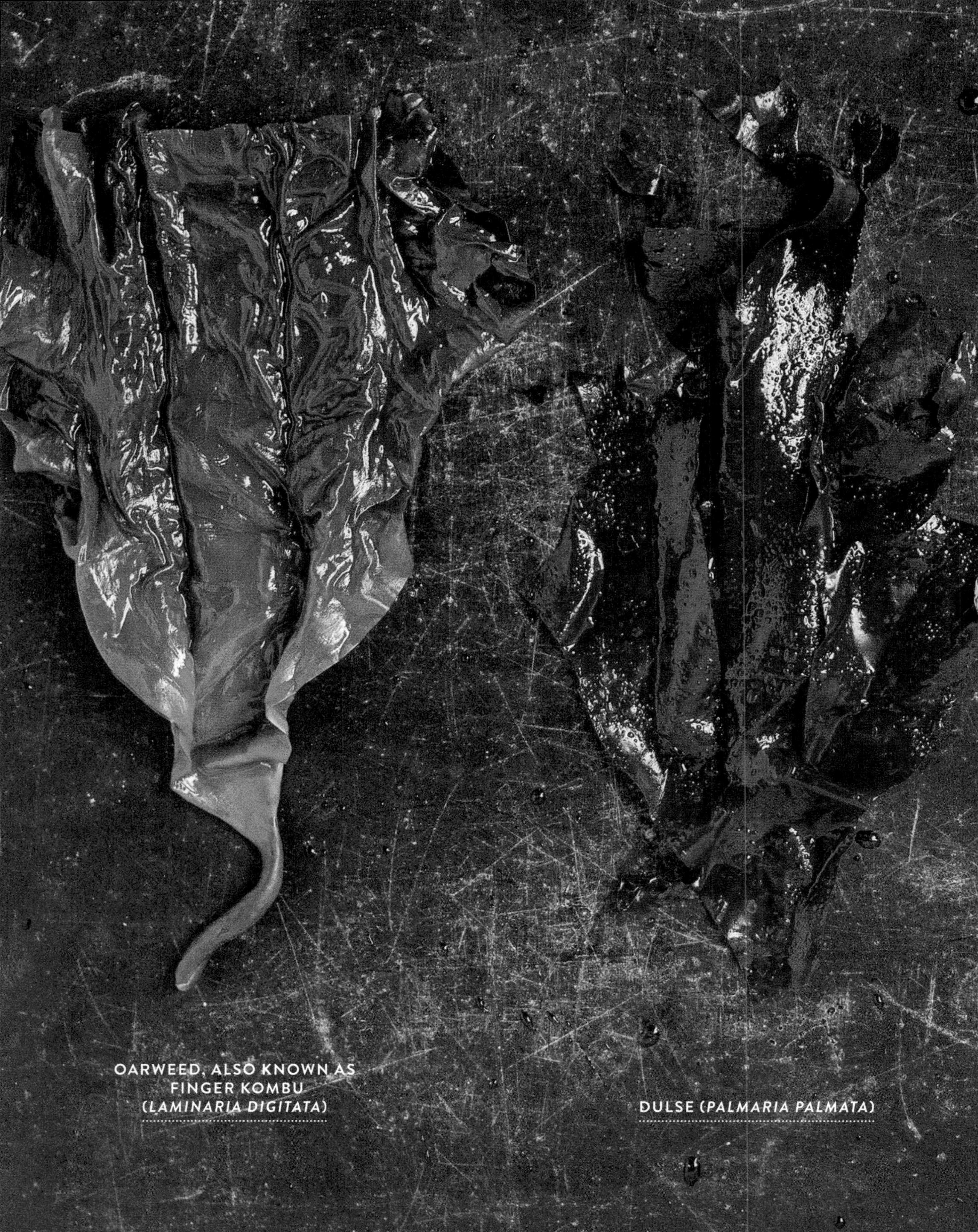

OARWEED, ALSO KNOWN AS
FINGER KOMBU
(LAMINARIA DIGITATA)

DULSE (PALMARIA PALMATA)

SEA LETTUCE (*ULVA LACTUCA*)   WAKAME (*UNDARIA PINNATIFIDA*)

SEAWEED: WHAT'S IN IT?

The composition and consumption of seaweed

IN ASIAN COUNTRIES, seaweed has been on the menu for thousands of years. Some researchers even believe that seaweed has been consumed in China since as far back as 2700 BCE. Furthermore, the Taiho Code, which dates back to 700 CE in Japan, stated that kelp and other algae were acceptable as payment for taxes to the emperor.

Seaweed has been used for centuries around the world, and not only in Asia. We know that seaweed was consumed in earlier times by the ancient Greeks as well as the indigenous populations of Hawaii, Scotland, and Scandinavia. The Romans used seaweed for joint disorders. In Ireland, people started collecting seaweed for consumption and to use as fertilizer around 1200 CE. The archeozoologist Ingrid Mainland has discovered that on the Scottish Orkney islands, seaweed has been fed to the local sheep for hundreds of years and that this tradition possibly dates back to the Neolithic era (the New Stone Age), about 5,000 years ago.

In a surprise discovery for phycologists, archeologists found a small basket with a "hash" of chewed *Gigartina skottsbergi* (a red marine alga) and other red algae mixed with the medicinal herb *boldo*. It was discovered in a "medicine shack" in Monte Verde, in the south of Chile. The mixture was used by the indigenous population, the Mapuche, as a medicinal tea that's good for stomachaches and colds. It's estimated that this seaweed basket is 12,500 years old.

Composition

What makes seaweed so special is that it works like a sponge: Seaweed absorbs nutrients directly from the water. The composition differs for each species of algae and is further affected by the temperature and nutrients in the water, the age of the alga, the amount of sunlight it is exposed to, the time of harvest, and the way it is stored.

Fresh seaweed consists mostly of water. Though the water content differs by species, it is always between 70 and 90 percent. In addition, seaweed contains macronutrients similar to sugars, proteins, and fats, and micronutrients such as minerals, vitamins, and antioxidants. The ratio between nutrients, liquid content, and caloric value likewise differs by species.

The caloric value of algae is relatively low, on average between 120 and 240 calories per 3.5 ounces (100 g) of dried seaweed. This, too, differs by species and depends on the method of conservation as well as the circumstances in which the algae are grown. Dried seaweed can, once soaked, increase its weight nearly tenfold, so 3.5 ounces (100 g) can become 2 pounds and 3 ounces (1 kg). In this chapter, we mean dried seaweed whenever we speak about nutritional value. Everything is calculated for the equivalent of 3.5 ounces (100 g) dried.

It's important to keep in mind that aside from healthy nutrients, seaweed can also absorb toxins from the water (e.g., PCBs and dioxins, heavy metals, or radioactive substances). The quantity of toxic substances in algae depends on several factors, including the quality of the water that they lived in, the time of harvest, and the processing. Furthermore, some seaweeds can themselves produce substances such as kainic acid and iodine, large amounts of which should not be consumed. So, it is important to buy seaweed that is safe for human consumption—consult your local producers.

We don't recommend consuming seaweed in unlimited quantities. Seaweed is better as a supplement to your regular diet. In Japan, where people are much more familiar with seaweed, they consume about 15 to 20 pounds (7 to 9 kg) per person per year, which is an average of 0.6 to 0.9 ounce (19 to 25 g) per day. For the average consumer, 0.2 to 0.35 ounce (5 to 10 g) of dried seaweed is a sensible supplement to a daily diet. Our advice is to eat a varied diet and add seaweed on a regular basis! Don't eat it only occasionally and then consume a large quantity all at once. (We have taken this into account for the recipes in this book!)

SEAWEED PROOF

In 2010, French researchers discovered that certain marine microorganisms are capable of breaking down specific carbohydrates that are found in red algae (*porphyran*, a carbohydrate from red algae in the *Porphyra*). The marine bacteria they studied contain an enzyme that is also found in bacteria of the intestinal flora in Japanese people. By transferring genetic material from one species of bacteria to another, the intestinal flora of these Japanese people is better capable of digesting seaweed.

Proteins

Proteins are important building blocks of the human body. Nearly all cells in our bodies contain proteins, but we're not capable of producing all of the ones we need ourselves. Here's the story: Proteins are specifically important for muscle growth, cell function, and many other processes. Other types of proteins, called enzymes, serve as catalysts to transform substances into other substances, as in the digestive system. And we have other types of proteins, such as antibodies. Proteins are made up of chains of amino acids. However, our bodies are not capable of producing some of these amino acids, the essential ones, which we must therefore take in via food.

The nice thing about seaweed is that it contains many of the essential amino acids. The percentage of protein differs by seaweed: Brown algae have the lowest concentrations, while red and green algae contain higher concentrations. For example, the most well-known brown alga in the North Sea is 5 to 10 percent protein. In comparison, the red alga dulse (*Palmaria palmata*) contains a high concentration of protein, sometimes up to 44 percent. The protein levels in green and red algae are comparable to those of protein-rich vegetables. So you see, seaweed is a great addition to any diet, plant-based or otherwise.

There is currently a lot of research investigating the extraction of proteins from seaweed. Someday, protein extracted from seaweed and purified may be used as a supplement in other products, like vegan protein bars or protein shakes.

Carbohydrates

The body needs carbohydrates to maintain good health. We use carbohydrates as fuel. They are necessary for the brain and red blood cells to function. When you take in carbohydrates, your body turns them into energy. Therefore, the type of carbohydrates and the amount you take in affect your health.

There are three kinds of carbohydrates in seaweed: digestible sugars (fast-acting carbohydrates), insoluble fiber, and soluble fiber. The percentage of digestible sugars in seaweed can run to 20 percent of its dry weight. This includes sugar alcohols like mannitol and sorbitol: natural sweeteners that are also found in various land plants. The carbohydrates in seaweed are the products of photosynthesis, which we mentioned earlier (see page 3). They are stored in the cell wall as a supply of energy and to support the cell structure by contributing to the solidity of the cell. Different species of seaweed contain different kinds and amounts of carbohydrates, and all of this also depends on the season and the location in which the seaweed grows.

DIETARY FIBER

Dietary fibers are indigestible carbohydrates. They are important for a functioning gastrointestinal (GI) system by creating the sensation of being full and helping to maintain a healthy weight. Some fibers help maintain a healthy level of cholesterol and thereby diminish the risk of cardiovascular diseases. Well-known carbs that the body cannot digest are pectin and cellulose.

Seaweed contains a lot of nutritional fiber: between 30 and 60 percent of its matter once dried. The way in which one of these nutritional fibers, phycocolloids (found only in algae and seaweed), interacts with the human body is remarkable. When you eat seaweed, the phycocolloids absorb water from your stomach and expand. This forms a gelatin-like structure, making you feel satisfied. Consuming these (soluble) fibers in food helps slow down high sugar levels in the blood. Other than that, phycocolloids are frequently used in food and cosmetics for their many properties, including as thickening agents, emulsifiers, stabilizers, or fortifiers. The three most well-known phycocolloids are alginates, agar, and carrageenan.

Omega-3 fatty acids are often found in fish oils. They are actually "algae fatty acids." It is often thought that fatty fish is the only source of omega-3 fatty acids. But fish don't make fatty acids; they take them in by eating microalgae (phytoplankton). Algae are therefore the original source of omega-3s! Nowadays, the fatty acids are being isolated from algae (farmed) and used for omega-3 supplements.

·······································

Fat

Fats are an important part of our diet. There are many different kinds of fat, though the main distinction is between unsaturated (healthy) and saturated (less healthy) fats. It is important to choose products with as little saturated fat as possible, to decrease the risk of cardiovascular disease.

Fats are used by your body as fuel, as building blocks for cells, and for transport. They provide a lot of energy: 1 gram of fat generates 9 calories of energy. That's much more than carbohydrates and proteins combined (4 calories per gram each). Aside from being a source of energy, fat is also a nutrient. It's an important part of your brain, nerves, and hormones. All body cells are partly composed of fat. Fat further carries and absorbs vitamins A, D, E, and K, which are fat soluble, throughout the body. Eating a sufficient amount of (good) fats and oils is therefore important for many internal processes.

The bioactive building blocks of fat are the fatty acids, many of which the body can produce from other fatty acids. As with amino acids, there are some essential fatty acids that the body itself cannot make and that need to be taken in through food, such as omega-3s and omega-6s. These fatty acids can be sourced from seaweed!

Seaweed contains relatively little fat, about 1 to 5 percent of its dry weight, though the amount differs per type of seaweed. Other factors include environmental circumstances and, more crucially, the season during which the seaweed grew.

The fatty acids in seaweed are mostly unsaturated and essential fatty acids. EPA (a type of omega-3) fatty acids can constitute up to 50 percent of the total fat content. These are important for brain function, among other things. Even processed seaweed (dried and preserved) still contains substantial quantities of omega-3 and omega-6 fatty acids.

Vitamins

Vitamins are organic substances that are essential for the body in small quantities. They play important roles in the development, recovery, and functioning of the body. Typically, we get vitamins through our food, since our bodies barely produce them—if at all.

Vitamins A, B1, B2, B3, B6, C, and E, in particular, are very clearly present in seaweed, but this differs per species and is dependent on conditions for development. It was once thought that algae formed a plant-based source of vitamin B12, but that isn't the case.

It should be noted that vitamin B12 doesn't appear in plant-based sources but is found mostly in animal products like dairy, eggs, meat, and fish. Seaweed does contain a vitamin B12–like substance, but recent studies have shown that this substance doesn't function in the human body, and therefore seaweed is not a source of vitamin B12. Because vitamin B12 is necessary to produce red blood cells and to help the nervous system function, people who follow a vegan diet are advised to supplement their diet with vitamin B12-enriched products.

Minerals and trace elements

Minerals and trace elements are essential for good health and normal growth and development. One instance is in the regulation of enzymes and hormones. Minerals are inorganic substances, and some, such as calcium and magnesium, appear in relatively large quantities in our bodies. Trace elements are minerals of which the body needs only a small amount, such as iodine and iron. For minerals, we're talking several grams a day; for trace elements, it's micro- or milligrams. Seaweed actively absorbs minerals and trace elements from salt water, and 10 to 55 percent of the dry material may contain them. The amount of minerals and trace elements in seaweed is therefore very high, from ten to one hundred times higher than that found in other common plant-based products.

To give you an idea: Corn contains about 2.6 percent minerals (dry material), and at 20 percent of its dry matter, spinach has an exceptionally high amount of minerals. Seaweed and algae can, depending on the circumstances, consist of 55 percent minerals: Iodine, calcium, phosphorus, magnesium, iron, sodium and potassium are present in especially high concentrations. Seaweeds have high concentrations of iodine. Some types contain a lot of iron (100 mg/100 g for sea lettuce), while others, like wakame, are rich in calcium (1,300 mg/100 g). Seaweed further contains a number of very important trace elements, like zinc, copper, manganese, selenium, molybdenum, and chromium.

MINERALS FOUND IN SEAWEED

Calcium helps with blood clotting, heart function, the nervous system, bone structure, and hormonal balance.
Phosphorus can be found in all cells. It's involved in all bodily processes, including developing bones and hormones, digestion, the making of proteins, and cell regeneration.
Magnesium stimulates enzymatic activities in cells.

IODINE IN SALT

In 1924, iodine was first added to salt in the United States because food didn't contain much iodine and many people were deficient. Since then, iodine has been added to table salt and low-sodium salt. The food label will indicate whether iodine has been added. Iodine is also present in high concentrations in seaweed.

Sodium helps with fluid balance, muscle function, and nerve impulses.

Potassium is good for fluid balance and blood pressure.

TRACE ELEMENTS FOUND IN SEAWEED

Iron is important for blood circulation and oxygen transport. It also helps produce red blood cells.

Zinc is a building block of many enzymes and hormones (such as insulin, growth hormone, and sex hormones), and it contributes to the functioning of the immune system.

Copper assists many enzymes; it is associated with digestion, the metabolism of proteins, and the formation of connective tissue, bones, and pigments for hair and skin.

Manganese is involved with the "energy centers" of the cell; it helps with the formation of bone tissue, detoxification, and the metabolism of proteins, carbohydrates, and fat.

Selenium, an antioxidant, works in detoxifying and binding heavy metals. It also protects red blood cells against free radicals, unstable and disruptive molecules in the cell.

Molybdenum is a building block of enzyme structures.

Chromium plays a role in carbohydrate metabolism.

Fluoride is necessary for the development and structure of the skeleton and teeth.

Iodine produces thyroid hormones and contributes to metabolism and growth processes.

IODINE

Iodine is an important trace element found in seaweed and has a large impact on our health. It is important for the regulation of the thyroid and is a building block of thyroid hormones. These hormones are important to many processes in the body, including metabolism and function of the nervous system. Consequently, a lack of iodine affects the production of thyroid hormones and damages the nervous system, most often in young children and pregnant women. People who don't live near salt water often don't get enough iodine, since it is almost exclusively found in saltwater products.

VITAMINS IN SEAWEED

Vitamin A, also called retinol, is good for your skin and helps create cells for the skin's tissue structure. The body is capable of making vitamin A from a plant-based provitamin A, like beta-carotene. It's needed for a good immune system.

Vitamin B1, also called thiamine, is crucial for the body's energy supply and is directly involved with the functioning of the heart muscle and nervous system. Vitamin B1 is also involved as a coenzyme with the many enzymatic processes in the body that are responsible for turning carbohydrates into energy.

Vitamin B2, also called riboflavin, functions as a coenzyme and plays an important role in drawing energy from carbohydrates, fats, and proteins.

Vitamin B3, also known as niacin, plays an important role in the production of fatty acids. Vitamin B3 is important in the metabolism of energy and protein.

Vitamin B6, also known as pyridoxine, is important for the breakdown and

CONTINUES →

construction of amino acids (the building blocks of proteins). Vitamin B6 regulates the working of certain hormones and is important for growth, blood production, the nervous system, and the immune system.

Vitamin C, also called ascorbic acid, functions as an antioxidant in the body (it protects cells against oxidation as a result of free radicals). Vitamin C is necessary for the absorption of iron from the digestive tract and is crucial for the immune system.

Vitamin E, also known as alpha-tocopherol, is fat-soluble and functions as an antioxidant. Vitamin E is important for the protection of cells, blood flow, and tissues. It is frequently used in skin-care creams.

Source: U.S. National Library of Medicine, MedlinePlus (nlm.nih.gov)

....

The thyroid is capable of storing iodine for years. This may allow a chronic lack of iodine in your diet to go undetected. Your body will live off its iodine storage until the supply runs out, after which you may start to experience symptoms: The thyroid will slow down and enlarge.

Seaweed is capable of storing iodine and has an iodine concentration up to 30,000 times higher than that of salt water. The best sources are brown algae. Half a gram of wakame is sufficient for the daily recommended dose of 150 micrograms. Children up to thirteen years old are advised to eat 120 micrograms per day. In Chinese culture, seaweed has been used for centuries to treat thyroid issues. In homeopathic medicine, doctors use brown algae to treat problems like obesity, bad digestion, persistent constipation, and flatulence.

Healthy people can consume more iodine, up to several milligrams daily. Too much iodine can lead to a hyperactive thyroid in people who are sensitive to it. This sometimes happens in countries with a traditionally high consumption of marine algae.

Seaweed and health

Because of its "magical" healing properties, seaweed used to be considered a luxury product. Today, it's an important part of Ayurvedic medicine (a Hindu healing system from India), thalassotherapy (saltwater therapy), phytotherapy (herbal medicine), and macrobiotic cuisine. Algae have been used as medication in China and Japan for hundreds of years, and seaweed was (and is) a substantial part of the daily diet and traditional herbal medicine in these countries. Seaweed is regarded in those regions as a treatment for tuberculosis, rheumatism, colds, open wounds, and intestinal worms. In recent years, seaweed has been promoted in various health movements and by "health gurus." It's seen as a superfood with many benefits to your health! It is presumed to help fight or prevent common ailments including thyroid issues, buildup of oxidized cholesterol, gastritis, arthritis, menopausal symptoms, skin problems, and esophagitis. Furthermore, seaweed—especially brown algae—contains antioxidants, which protect against cardiovascular diseases and some forms of cancer. Because of the potential benefits of antioxidants, research into the positive effects of seaweed has gotten a boost. It is, however, important to realize that these health claims have not been sufficiently examined and are often unproven.

A few notable observations: In areas where the population has a diet rich in seaweed and other saltwater products, there are fewer occurrences of cardiovascular disease and high blood pressure. In 1927 the Japanese professor Shoji Kondo of Tohoku University suggested there is a correlation between areas in Japan where a lot of seaweed is consumed (the coastal areas) and a higher life expectancy—especially for women. These people ate less food anyway and consumed less salt. Seaweed contains significantly more fiber than fruit and vegetables and makes you feel satisfied sooner. It also has an umami flavor (see page 20); because it's so flavorful, by adding seaweed to their diet, people typically use less salt and eat smaller portions.

The health benefits of seaweed consumption are mostly based on old traditions, writings, and experiences. To date, research has focused on the connection between seaweed consumption and lowering the risk of cardiovascular diseases and cancer.

There are a few important ways in which seaweed contributes to a healthy diet: It contains many minerals and vitamins that can easily be absorbed by the body. Additionally, because of the high concentration of fiber in seaweed, sugars in the digestive system are absorbed more slowly, which causes blood sugar levels to rise at a slower rate (the same effect that whole wheat flour has). Finally, seaweed contains many healthy fatty acids and essential amino acids.

VITAMINS & MINERALS

Vitamins and minerals have similar functions, and both are needed for important bodily processes. We absorb them through food and water. The most important difference between vitamins and minerals is chemical: Vitamins are organic compounds and are made by certain plants and animals. They are organic substances that contain hydrogen and carbon. Minerals and trace elements are inorganic compounds; they come from rocks, ore, or soil. Minerals and trace elements slowly dissolve in water and can be absorbed by plants and algae. Only then can animals and humans take them in via food and water.

THE TASTE OF SEAWEED

WITH A RICH VARIETY OF seaweed species to choose from, there's a wondrous world of flavors and culinary possibilities waiting to be discovered by the seaweed consumer. Most species are edible, but not all algae are equally suitable for the novice 'weed-eater. Typically, the most edible seaweeds are the ones that live in salt water. Freshwater algae are much less suitable for consumption. You will perceive a big difference in taste among different kinds of seaweed.

Oarweed, for example, contains strong and salty flavors, while sea lettuce is soft with a refined taste to it. If you have no experience eating seaweed, it's best to begin with an accessible seaweed species that is easy to use in combination with other ingredients. Given the wide variation in taste and texture, it's an adventure to find out what you like and with which dishes they go well. Some algae can be eaten raw; all you have to do is just rinse them with water. Most algae are best cooked, dried, baked, or roasted—and they taste wonderful.

The texture of seaweed

Besides for its unique taste, seaweed is valued by fans for its texture and mouthfeel. Seaweed comes in various different forms, from powder or roasted leaves to chunks or spaghetti. For most people, the new flavors and textures take a little getting used to. Some species taste great in dried form, while others are best when fresh. Dried seaweed will spring back to its old form when soaked in water. Some fresh or soaked seaweeds have an almost rubbery texture; others are very tender or somewhat crunchy. Certain species are more suited for cooking, braising, or roasting. Roasted seaweed is often crunchy and crispy, and it melts in your mouth: perfect for sprinkling over a salad. So that you can learn to appreciate the taste, we advise you take it easy at first by adding small amounts to your favorite dishes.

The taste of seaweed: Umami

For centuries, seaweed has been consumed because of its nutritional value. The intense flavor, however, makes seaweed much more than just the nori around your sushi or the wakame in your ramen. There are many more wonderful seaweeds to experiment with, and each kind has its own unique taste. What all algae have in common is a distinct umami taste. Because umami differs significantly from sweetness, sourness, saltiness, and bitterness, it is commonly called the fifth basic taste. Thanks to the worldwide introduction of Japanese and other Asian cuisine, umami is increasingly finding a platform in the West. For a long time, umami was characterized as "hearty," "brothy," or "meaty," but that's mostly due to the lack of a better term. The word *umami* is Japanese in origin and literally means "pleasant savory taste."

In 1908 the Japanese professor Kikunae Ikeda discovered the flavor-enhancing properties of glutamate (a saltlike substance). Ikeda suggested referring to the flavor sensation of glutamate by the term *umami*. Glutamate is a nonessential amino acid that occurs naturally in meat, fish, dairy, breast milk, vegetables, yeast extracts, mushrooms, soy sauce, fermented soy products (such as miso), and various cheeses (such as Parmesan). Seaweed also naturally contains glutamate. The novice seaweed eater tends to find it difficult to describe the flavor of seaweed: One person will say "fishy," while another will say "salty," and a third calls it "briny." Of course the taste of seaweed depends on the species. Taste factors include age (even within the same species), how it has been prepared, and its origin. Seaweed that grew in a relatively salty body of water, in the Mediterranean for instance, has an even more pronounced umami taste. It was scientifically proven only recently, in 2000, that humans have taste receptors for umami. Our tongues have special umami receptors!

SPIRULINA & CHLORELLA

Spirulina and chlorella are tiny unicellular organisms. Not just food supplements, these algae are an abundant source of chlorophyll. Plants use chlorophyll for photosynthesis: converting light energy from the sun into chemical energy stored as carbohydrates. Chlorella is a *freshwater* algae with very strong cell walls. These cell walls need to be pulverized before our bodies can access the valuable nutrients stored inside. Spirulina, on the other hand, is a *saltwater* algae. These superfoods contain high concentrations of proteins, vitamins, and minerals.

As we've already discussed, certain seaweed species also contain a high quantity of iodine, which can result in a strongly dominant salty flavor. Don't forget that seaweed adds a natural form of salt to a dish; it's less strong than sea salt but definitely influences the flavor and overall salt content. You should always taste a seaweed dish before seasoning it further. In many cases, you can even omit the salt all together.

Get to know your seaweed. After getting used to this ingredient and experimenting with it, you will notice that seaweed doesn't taste as exotic as it initially seemed. You may be new to the term *umami*, but the flavor will be familiar to you from many everyday products, such as ketchup. The only difference is that you now experience umami in a very direct and pure way: It is the essence of seaweed.

Different types of seaweed in the kitchen

There are many types of algae used in recipes, most notably in Asian cuisine. Among the most common types are dulse, Irish moss, nori, kombu (kelp), wakame, arame, hijiki, toothed wrack, fucus, devil's apron, Norwegian kelp (or knotted wrack), oarweed, sea lettuce, and sea spaghetti. All of these are deliciously salty and briny, yet each has its own distinct character. Here we will describe some of the most commonly eaten algae.

Brown algae

KELP (KOMBU)

Kelp is a collective name for approximately three hundred different species of brown algae belonging to the order of Laminariales. These brown algae grow close to the surface in coastal waters and near rock plateaus. They thrive in colder, more sheltered seas and oceans in places along the Atlantic and Pacific coasts of the United States and the coasts of Europe and Japan. The more well-known species of kelp include *Laminaria digitata*, known as oarweed or finger kombu; and *Saccharina latissima*, or devil's apron, also commonly called sweet kombu.

Kombu (*Saccharina japonica*; *konbu* in Japanese) is the most well-known brown algae and is sometimes called the "King of the Sea." This kombu is used a lot in Japanese, Chinese, and Korean cuisines and can be eaten in several forms: dried, fresh, roasted, frozen, cooked, stir-fried, marinated, candied, or even as sashimi. It can also be used to make tea. The most famous use, however, is as one of the main ingredients for fresh dashi bouillon or stock.

As with most seaweed, the taste and texture of kelp varies depending on species and age. Because of its high glutamate content (see page 21), the versatile kelp is known for its deep umami flavor. It is also an excellent salt substitute with a familiar, accessible sweet-salty flavor. Kelp species are rich in many minerals, including calcium, potassium, magnesium, and iron, and the trace elements manganese, zinc, chrome, and copper.

WAKAME

Wakame means "young girl" in Japanese. The alga was named after the young Japanese seaweed gatherers and after the graceful way in which it dances in the current.

Wakame (*Undaria pinnatifida*) is a brown algae that is native to Asian waters, where it grows near rough, rocky coastlines. The plant lives 20 to 40 feet (6 to 12 m) below the surface and can reach a length of 24 to 48 inches (60 to 120 cm) and a width of about 16 inches (40 cm). It ranges in color from olive green to brown. The individual leaves are rubbery, long, and smooth and have wavy edges. It has a small, highly branched holdfast.

DEVIL'S APRON
(*SACCHARINA LATISSIMA*)

DULSE (*PALMARIA PALMATA*)

WAKAME
(*UNDARIA PINNATIFIDA*)

SEA LETTUCE (*ULVA LACTUCA*)

KOMBU TEA

Kombu tea is very popular in Japan and is available at most Asian markets and organic stores in the United States. It has a high iodine content, which can help stimulate metabolism and the thyroid. Sushi rice is often cooked in water that kombu has been steeping in for an hour or so.

Wakame has lower iodine content than other brown algae and is high in calcium.

The taste of wakame is often described as soft and sweet, with a fine texture. It's a key ingredient for miso soup and is often used in salads. In Japan, wakame is seen as a delicacy and served fresh, presoaked, marinated, or processed. Raw and marinated wakame are particularly delicious in salads and added to vegetable dishes or stir-fries.

Wakame is widely available dried and needs to be soaked only briefly. Take into account that soaked wakame increases in volume about tenfold.

HIJIKI (HIZIKI)

Hijiki (*Sargassum fusiforme*) is a brown alga that is native to the Pacific coasts of Japan, China, South America, and Hawaii. Like wakame, hijiki has been a staple in Asian cuisine for centuries.

Right after it's harvested, hijiki is washed, cooked, and dried. Hijiki ranges in color from green to brown and often has scores of very small leaves only 1 to 2 inches (2.5 to 5 cm) long. Dried hijiki looks like tea leaves of ½ to 1½ inches (1 to 4 cm) in size. For novice algae-eaters, this seaweed may be a bit of a challenge, but for the connoisseur, its very pronounced umami taste is fantastic. Dried hijiki has a relatively salty flavor.

Hijiki is harvested in the spring. After the harvest, hijiki is dried in the sun, cooked in water, and then dried again. In Asian cuisine it is served as a side to rice, and it is especially delicious when paired with soy sauce! Hijiki is rich in dietary fiber, iron, calcium, and magnesium, but don't eat too much of it. This seaweed tends to store higher levels of inorganic arsenic, which at high concentrations is poisonous for humans. Because of this, food agencies in America, Canada, and Britain have advised against eating it. If you would like to try hijiki, always be sure to buy a variety that has been grown in uncontaminated waters and is safe for human consumption, and be careful not to consume more than 2 tablespoons of hijiki per week. For more information, we recommend consulting edenfoods .com/hiziki.

ARAME

Arame (*Eisenia arborea, Eisenia bicyclis*) is a brown alga that grows on both sides of the Pacific Ocean—in Japan as well as in South America. It lives on open rocky coasts where there's a big tidal range (think: ebb and flow) and prefers cold waters. To dry arame after it has been harvested, it is laid out in the sun and then cooked and cut into strips, a process comparable to that used for hijiki. The color of dried arame is very dark, almost black. When arame is soaked in water, it turns brown and significantly increases in volume. Don't soak it for more than five minutes; otherwise it will lose its taste. Dried arame—as sold in stores— needs to be cooked for only a couple of minutes after soaking. It can also be steamed or fried in a wok as part of a stir-fry. Arame contains a lot of calcium, iron, magnesium, and vitamins A and B3. Because of its mild and sweet flavor, Arame is commonly considered to be the sweetest of seaweeds. It has a very subtle taste and is therefore easy to use in a variety of dishes, from sushi, soup, and salad to savory muffins and stews.

SEA SPAGHETTI

Sea spaghetti (*Himanthalia elongata*), also known as thongweed, is another brown alga. It grows on the Irish, British, and French coasts and can also be found in the Baltic Sea. The alga is known for its elongated green-brown cords, some of which can be many feet in length. The cords spring from the weed's holdfast, which generally attaches itself to boulders or shells. The alga consists of the holdfast, a buttonlike thallus (the body of the algae), and the long cords, which carry the reproductive organs. These cords can grow remarkably fast and are best harvested in spring, when they are at their most delicious. Sea spaghetti can be enjoyed raw. It has a smooth, salty flavor. Dried sea spaghetti first needs to be soaked in water for fifteen minutes and then blanched for five minutes if necessary. Rich in dietary fibers, vitamins A, C, and E, and iron, sea spaghetti makes for a great grain substitute in pasta dishes.

Red algae

DULSE

Dulse (*Palmaria palmata*) is a red alga that grows predominantly on the northern coasts of the Atlantic and Pacific oceans. The *P. palmata* is the only *Palmaria* that can be found on the European Atlantic, namely near Portugal, Iceland, and the Faroe Islands. It also grows near Russia, Canada, and Alaska, as well as on many Asian coasts.

Dulse has a deep red-brown-purple color and flat leaves that grow directly on the holdfast, making it look like a breezy palm tree. Initially, dulse consists of a single flat leaf, but as it gradually widens it branches out into broad segments of about 20 inches (50 cm) in length and 1 to 3 inches (3 to 8 cm) in width. Dulse attaches itself to rocks or other seaweed species (such as kelp). The leaves have a distinctive leatherlike texture.

The seaweed is harvested during the end of the summer and can be picked by hand. Although it can be eaten fresh, dulse is usually dried. For dulse to dry, it needs to be rinsed first and then laid out in the sun for quite some time. While it dries, the salt stored inside the

DULSE (*PALMARIA PALMATA*)

alga can become visible on its surface in the form of white spots. Note that dulse shouldn't be cooked long, because it will fall apart.

Dulse can be prepared in a pan (fried) or in the oven. It is often added to soups, stews, sandwiches, and salads, or used as an ingredient to infuse pizza or bread dough with some sea flavor. Dulse makes for a great salad topping, too. With its soft and crunchy bite, it is the perfect weed for the true seaweed-snack addict. Dulse contains many dietary fibers and has a nutty, salty-sweet flavor. It also has relatively high protein content—higher than that of chicken or almonds.

NORI (PURPLE SEAWEED)

Nori (*Porphyra umbilical, Porphyra tenera, Porphyra yezoensis*), also known as purple laver, is the most well-known and the most frequently used seaweed in the world. Nori, whose name literally means "alga" in Japanese, is in fact a combination of various red algae. Because nori sheets consist of a combination of *Porphyra* species, their color and quality may vary. The cheapest sheets sell for a couple of cents each; the most expensive ones cost tens of dollars. Nori can be found on the coasts of Europe, the United States, the Philippines, and Japan, where it grows in abundance. Nori is harvested from the beginning of spring until summer. In Japan, to enable new growth, nori is picked carefully. It is then rinsed and dried in the sun or in ovens. It is only after drying that nori gets its black-purple color. Nori is sold dried as well as roasted, which requires the leaves to be dried and finely ground immediately after being harvested. The remaining pulp is then spread out, dried, and roasted. The roasting is done at 200°F (95°C), until the seaweed is crispy.

The familiar green color appears only after roasting, after which the nori is cut into sheets of about 8 inches (20 cm). Nori can be used in many ways: Most famously, it is used for sushi rolls, but it is also delicious with beans, vegetables, cereals, pasta, and rice. It's also scrumptious as a snack and in salads, stir-fries, stews, and soups. Nori has a strong sealike flavor and high protein content. It contains a lot of iodine, carotene, and vitamins A, B, and C.

IRISH MOSS

Irish moss (*Chondrus crispus*) is a red alga native to the Atlantic coasts of Europe and North America. The alga is about 8 inches (20 cm) tall and feels spongy. This tiny seaweed is primarily known as a source of carrageenan. More than 50 percent of its weight can be made up of this unique carbohydrate. When processed, carrageenan has the ability to retain a lot of water. It is therefore used as a stabilizer or thickener in many foods.

Green algae

SEA LETTUCE

Sea lettuce (*Ulva lactuca*) is a green alga that is native to the coastal zones of all the world's oceans. It thrives in nutrient-rich environments. Though this alga is abundant during the summer, it grows all year round. The leaves are very thin—only two cell layers thick—but still firm. The smooth blade can be round but also very irregular in shape, with all sorts of lobes forming. Sea lettuce, as the name suggests, looks more or less like "land" lettuce.

Sea lettuce has a short stem that usually attaches itself to rocks or other solid surfaces. The remarkable thing is that even when it's floating in the sea—say, after a strong storm— the alga just continues to grow. Because sea lettuce is an annual plant, it dies off around the same time, around autumn: That is when a lot of sea lettuce can be found on the shoreline.

In France, many tons of sea lettuce wash ashore each year. For generations, this seaweed has been used as a natural fertilizer.

Sea lettuce contains a lot of magnesium, calcium, vitamin A, and vitamin C. It doesn't contain as much iodine as other types do but does have a considerable amount of iron. It is relatively high in protein and low in fat. Sea lettuce is often eaten raw in salads. It is also used as a flavoring in soups or cooked with other vegetables. It is perfectly suited for steaming, baking, or making tea. It's crispy, soft, and fresh with an almost nutty, salty-sweet spinach flavor. Dried sea lettuce is fairly firm, but after it is soaked in water for a few minutes, it becomes soft. You can cook it in just five minutes!

Harvesting fresh seaweed

Fresh seaweed may be difficult to find in the United States. Organic specialty stores and online sellers offer only a limited selection of edible algae. The reason is that fresh seaweed is highly fragile, and soon after it's been harvested, it begins to break down. Seaweed that has been cut and no longer floats in seawater dies off fairly quickly. In this way, it's like a flower: Once cut, it can only survive for a short while, provided it gets enough water and nutrients. The texture and color of certain algae will start to change after they've been out of the water for an hour. Other saltwater algae will become slimy and start to degrade as soon as they come into contact with freshwater (so be careful when you rinse them!). To combat degradation, seaweed is often processed directly after being harvested. The algae are rinsed with clean (salt) water and subsequently dried in a dryer or in the sun. Once dried, the algae are vacuum-sealed. Their color and flavor will still change slightly under the influence of such elements as moisture and light.

Brown algae are the strongest of all seaweed varieties. They contain organic compounds that naturally protect the algae against bacteria and fungi. On top of that, brown algae are better able to withstand moisture from the air. They are therefore often not packaged with a completely airtight seal, which may cause the algae to develop a mysterious white coating. This is glutamate, the component of seaweed that lends it its delicious umami taste (see page 20). Be careful not to scrape off this layer, or that wonderful taste will be lost! When you buy dried seaweed from the store, it may need to be rehydrated by submerging it in water before use. Some algae require a couple of minutes, others at least an hour. Rehydrated seaweed keeps in the fridge for one day. Fresh, undried seaweed is often pickled directly after harvesting, in the same process by which vegetables are often preserved. Preserved seaweed needs to be soaked in ample water to wash off excess salt. Afterward, rinse it and gently squeeze the water out. This type will keep in the fridge for one day and is also ready for immediate use.

DEVIL'S APRON
(SACCHARINA LATISSIMA)

Seasoning a dish with seaweed

Seaweed is an excellent salt substitute. Salt has been used to enhance the flavor of our food for centuries. Table salt is mainly composed of sodium chloride. Salt consumption in the United States is too high, and therefore sodium intake is too high. Although seaweed also contains salt, part of its makeup is another type of salt, potassium chloride. This makes seaweed a perfect salt substitute, and the fact that it contains less sodium is a nice plus.

THE CULTIVATION & HARVEST OF SEAWEED

LONG AGO, inhabitants of coastal regions gathered their seaweed directly out of the water or from the beach. Seaweed that grew near the coastline or that washed ashore after a storm was used by the locals for food or as fertilizer. We assume that the cultivation of seaweed started in Japan between the fifteenth and sixteenth centuries. There are differing stories and theories about how and when this occurred. One of the stories has it that a shogun, a military commander, ordered the fishermen in the small village of Shinagawa (now Tokyo) to provide him with fresh fish daily. The fishermen decided to build a cage off the coast in which to breed the fish. Over time, seaweed began to grow in the cage, which the fishermen then started cultivating as well. Sources report that as early as 1670, seaweed was being cultivated in Tokyo Bay. They threw bamboo in shallow water, which attracted seaweed spores; from these spores grew seaweed. Much later, during the mid-twentieth century, seaweed farmers discovered that the spores actually adhered much better to synthetic materials. Since then, synthetic nets have been tied to the bamboo stalks, thereby increasing the harvests.

In Europe, seaweed was mainly harvested in the wild. One of the first mentions of seaweed harvest was found in a French text from 1681, in an ordinance stating the harvest seasons and the number of authorized harvesting days. The first document referring to the commercial use of seaweed also dates back to the seventeenth century.

The population on the coasts of Brittany burned seaweed and used its ashes for the production of soap and glass. Beginning in the nineteenth century, seaweed was increasingly used for the production of iodine. Until World War II, this was its primary commercial use. After the war, with the rise of the chemical industry, the seaweed industry declined. Seaweed was then harvested mainly to obtain phycocolloids, such as the thickener alginate.

For all those centuries, seaweed was harvested manually. At low tide, it was cut with knives. After a big storm, the shores were lined with washed-up seaweed, which was then collected by the locals, using knives, rakes, pitchforks, sickles, and nets. Seaweed that grew farther from the coast was also collected from kelp forests. Seaweed farmers would cut the

seaweed from a boat. In the 1970s, seaweed foragers in Norway, France, and Spain began harvesting mechanically in the wild. This development was due to a large demand for alginate.

Market

Seaweed is grown in many countries—sometimes on a small scale for local use and sometimes on a large scale for export. According to the Food and Agriculture Organization (FAO) of the United Nations, China, Indonesia, the Philippines, Korea, and Japan account for about 95 percent of total seaweed production. More than 53 percent of all seaweed comes from China. Indonesia produces 26 percent, the Philippines 7 percent, and Korea and Japan share the remaining 9 percent. The rest of the world accounts for only 5 percent of seaweed production. In recent years, this has begun to change: Asian countries have a growing interest in European algae. Because European waters are cleaner, the algae are perceived to be safer for consumption.

Seaweed production doubled between 2000 and 2012 globally. In Indonesia especially, the market has grown considerably. The FAO expects that this growth will continue in the coming years. FAO statistics show that a total of thirty-seven species of seaweed are being cultivated worldwide. The most-cultivated seaweed varieties are Japanese kelp, wakame, nori, and various species of red algae, such as *Eucheuma* and *Gracilaria* species, which are used for the production of agar and carrageenan.

From World War II until recently, Europe saw a downturn in seaweed production, though after the turn of the century, interest increased. Recent years have seen a rise in (small) companies that produce, cultivate, and sell algae (online). The consumer market is relatively small but clearly on the rise. Europe has an advantage over Asian countries in terms of the quality of the product. The biggest seaweed-producing countries in Europe are Norway, Ireland, and France, and one of the world's largest carrageenan and alginate producers is located in Denmark. This company sources most of its algae from the West but also imports a lot from Asia.

Europe has a big market for alginate, a special carbohydrate extracted from seaweed, which is used as a thickener. One obvious reason why the production of seaweed in Europe is so much smaller than in Asia is the fact that seaweed consumption is significantly lower. A second reason is that cultivation and harvest are labor intensive and therefore expensive. Consequently, European initiatives focus on mechanized cultivation and harvesting methods of high-quality algae. Up until now, most of the seaweed harvested in Europe came from natural populations, that is to say, from wild harvest. When the seaweed is collected in a sustainable way, using a specialized crane to gather it from the sea, it will grow back and can continue to reproduce.

The most harvested algae in Europe—primarily kelp like oarweed and knotted wrack—are species that grow mainly in forests, which contribute to ecological diversity. Keeping this in mind, seaweed farmers must take precautions to harvest correctly in order to not harm these ecosystems.

Seaweed harvesting: Wild harvest or cultivated crops?

There are several methods for harvesting seaweed, so it is useful to differentiate two main classifications: seaweeds from a natural population (wild harvest) versus cultivated (farmed) seaweeds. Wild harvest simply means the algae grew naturally before being harvested; there was no cultivation. The opposite is true for farmed seaweed: Seaweed farmers implanted seedlings in cultivation systems (farms).

In farming, we can roughly distinguish between two methods. One is the collection of spores or gametes, which can be found in areas where many types of seaweed grow. Nets are placed in these areas, attracting spores to attach themselves to the nets. The collected spores are then fertilized, allowing them to grow into new adult algae. This method allows farmers to increase the natural growth rate of the algae. It is often used by seaweed farmers in Asian countries.

In addition to this "spore catching," it is possible to grow algae in a protected and controlled environment. For this method, seedlings are first grown on a thin rope. Once they become large enough, the rope is then wrapped around a thicker rope. Another method is spraying a gel containing algae seedlings directly onto a rope, cloth, or netting. These materials are then placed in open sea or in a bay where the seedlings grasping only a few millimeters of rope, cloth, or netting can eventually grow into seaweeds several feet tall.

Finally, some farmers take advantage of algae's natural vegetative reproduction. Because algae reproduces on its own, it's possible to pull cuttings from some seaweeds (like sea lettuce) to be regrown elsewhere. A piece of a parent alga is removed and then cultivated to grow back to a full-sized seaweed at another location.

Farming systems and locations

The systems used for the cultivation of seaweed can be divided into three types of locations: onshore, inshore, and offshore seaweed farms. An onshore farm is situated on land. An inshore (sometimes nearshore) seaweed farm is close to or just off the coast. Here, the algae are either grown in ocean water, a bay, or an estuary such as a fjord. Offshore farming describes the process of growing algae in the open sea.

ONSHORE (LAND)

Onshore cultivation is used for more sensitive or specific seaweed species as well as at research facilities. This technique involves large water-filled tanks, sometimes excavated, in which the algae are grown. It often takes place near the coast so that water can be pumped in directly from the sea. Tanks like these offer the seaweed a calm, protected environment; as long as there is sufficient sunlight and a fresh influx of seawater, the algae can prosper. This practice of large-scale cultivation is particularly popular in South America. Another advantage is the level of control it offers: There is less chance of contamination and growth of microalgae. Inspection and harvesting are much simpler, too. The disadvantage is that, partly due to the pumping in of seawater, this production method is costly.

INSHORE OR NEARSHORE (COASTAL)

Almost all seaweed sold in stores comes from coastal areas. At the coast, you'll find both naturally grown algae and cultivated seaweed. Almost all the seaweed in today's world grows nearshore. This is one of the simplest methods to cultivate seaweed. The sea provides the nutrients, the waves are less powerful, and the algae are relatively accessible to farmers, which allows for easier control and harvesting of the product.

There are some requirements a coastal zone must meet in order for the algae to thrive there. First, a lot of algae live in salt water and cannot survive in freshwater. It is therefore important that the bay or the coastal area is not near an estuary, which receives freshwater runoff from land. Furthermore, the water should be frequently refreshed (through currents or tides) so it's clean and nutritious. Finally, most algae can't live without water for too long, so it's important that an area doesn't fall dry for too long.

OFFSHORE

Although offshore seaweed cultivation isn't yet common practice, it does have some clear advantages, including ample room for scaling up. Integrating several offshore activities has many possible benefits, including shared logistic responsibility and increased resources provided by a shared ecosystem. And water in open sea tends to be cleaner because of the high refreshment rate created by waves and currents. Innovative small-scale pilot studies with offshore seaweed farming are currently being done in the Dutch North Sea, and the results are encouraging. Large-scale seaweed farming on open sea could

FORAGING FOR WILD SEAWEED

Foraging your own seaweed is something for the well-prepared adventurist in possession of sufficient knowledge of edible algae and clean bodies of water. Picking the wrong seaweed can lead to food poisoning, so be careful when foraging by yourself or join a qualified professional seaweed harvester. It's fun *and* informative! For more on foraging seaweed, see resources on page 166.

greatly impact the way we source our food—from the sea rather than from land. We already have offshore wind farms; now think about using that vast ocean acreage for cultivating seaweed, too. As offshore, large-scale farming develops, people will likely focus more on the commercial value of seaweed's organic components and less on its production for food.

Applications outside of the kitchen

Although seaweed and salty plants are wonderful seasonings in the kitchen and greatly enhance vegetables, seaweed has many more applications. For example, algae are sometimes used as feed (or feed supplement) for horses and other animals. They're also used as an ingredient for making soap. Seaweed is a source of protein that can be used to enrich food products. The same applies to phycocolloids like agar, carrageenan, or alginate, which are used as natural thickeners (see page 50). Additionally, seaweed contains mannitol and sorbitol, substances that are used as natural sweeteners.

Seaweed contains nutrients (nitrogen, phosphates, and other minerals) that are essential for the growth of plants. Making use of the phosphates and minerals in seaweed could create a closed nutrient cycle: The nutrients (nitrates and phosphates) from wastewater that end up in the marine habitat are absorbed by algae, which can then function as fertilizer to grow new crops.

Seaweed and algae are also used in thalassotherapy. Thalassotherapy, from the Greek words *thalassa* ("sea") and *therapy* ("treatment"), is any form of treatment that involves the sea, including the use of seawater, sea air, algae, minerals, sand, clay, plankton, or shells. People with skin disorders or skin problems use thalassotherapy, and ingredients from the sea are used in many skin-care products: algae masks, anticellulite algae packs, and salt baths, to name a few. As early as Roman times, this therapy was prescribed to patients with rheumatoid arthritis and psoriasis. The properties of seawater—minerals and organic materials such as algae—are believed to be beneficial to your health. Pharmaceutical, cosmetic, and food companies also use essential ingredients from seaweed in their products: Body lotions, face creams, mascara, and toothpaste all contain seaweed-derived ingredients. Seaweed is used in the production of paper and textiles, and research is being done into the use of algae as a biofuel.

Environment

Algae are found in all of the world's oceans. They are essential for life in our seas and beyond, often serving as a food source for fish and other sea creatures. Certain species of algae (like kelp) form dense forests where fish, crab, and other marine animals find shelter from waves and currents. These underwater sanctuaries are ideal hiding places and are often used as a breeding ground and nursery for the ocean's offspring. It is therefore important to proceed with caution when harvesting seaweed from a kelp forest.

Growing algae in seaweed farms is the better option for the sustainable production of nutrients; it has less impact on the environment. The beauty of seaweed is that its cultivation requires only resources that are present in the ocean, and it doesn't depend, like agricultural crops do, on freshwater and fertilizers.

Seaweed can also be used to purify water with too-high concentrations of nitrogen and phosphorus caused by fertilizers. The algae convert these nutrients, which in high concentrations are harmful to the ecosystem, into plain biomass. By removing these surpluses from the water, the ecological equilibrium is restored.

OMNIS VITA EX MARE:
ALL LIFE ORIGINATES FROM THE SEA.

SAMPHIRE
(SALICORNIA)

SEA ASTER
(*ASTER TRIPOLIUM*)

COMMON ICE PLANT
(*MESEMBRYANTHEMUM
CRYSTALLINUM*)

SEA VEGETABLES

SEA VEGETABLES owe their distinctive flavors to their relationship with the ocean and seawater. These are plants that grow on land but have a strong resistance to salt water. These marine plants are called halophytes and are related to some more well-known "land vegetables": Sea aster is similar to lettuce, chicory, and black salsify; marsh samphire or "sea fennel" resembles ordinary fennel; and sea kale is part of the cabbage family. These plants live on the border between land and sea. Their great similarity with seaweed is that they contain many vitamins and minerals from the ocean that are can easily be taken in by our bodies. Below you'll find a number of known and lesser-known delicious sea plants.

Samphire

Samphire (*Salicornia*) is a salt-tolerant, annual plant of the amaranth family (Amaranthaceae). Similar to cactus, samphire is a succulent, a plant type better resistant to drought than other land plants. It is native to the United States and Europe, where it grows in salt marshes, on beaches, and among mangroves. Often it is the first plant that pops up on new stretches of salty soil, especially in delicate habitats that are often inundated with seawater, such as salt marshes. Samphire is a small green plant, often less than 8 inches (20 cm) tall, with a strong stem and straight branches. The nearly invisible leaves cover the stem and branches and look more like scales.

Samphire is also called glasswort, pickleweed, and marsh samphire. In the United States, this plant grows from spring to fall. This vegetable has a relatively high protein and salt content. The latter makes it necessary to thoroughly wash the plant before use, in order to "discharge" some of the salt from within.

In terms of taste, samphire has a pronounced salty flavor. It is also easy to prepare and can be eaten raw, stir-fried, cooked, or blanched. Its taste and texture are sometimes compared to that of wild baby spinach.

Sea aster

Aster (*Aster tripolium*), often called sea aster, is a plant belonging to the sunflower family (Asteraceae or Compositae). This species lives around the high-tide line. Sea aster is a

6- to 24-inch (15 to 60 cm) biennial plant with oblong leaves that are about 5 inches (12 cm) long, and it blooms in the late spring and early summer. This plant lives on soils with a relatively high salt content and has glands to excrete salt. It can be found in mostly coastal areas in North Africa, China, Europe, and eastern Russia, and is not native to the United States. In autumn, sea aster changes color from green to red.

The nickname *lamb's ear* comes from the long green leaves that are shaped like a lamb's ear. The young leaves are eaten as a vegetable and sold under the name of sea aster. Aster grows in tidal areas such as the Dutch Wadden Sea—a UNESCO World Heritage Site. These days, the vegetable is cultivated in these areas as well as harvested in the wild. Spring and early summer are the best times for harvesting.

As with most sea vegetables, sea aster has a salty taste: Briefly rinse the leaves to reduce the salt, but not for too long, because the plant isn't very tolerant of freshwater.

Oyster leaf

Oyster leaf, also called sea bluebells or oyster plant (*Mertensia maritima*), grows along the beaches and coastlines of North America and Northern Europe. Part of the borage family (Boraginaceae), oyster leaf has a distinct appearance. Inhabitants of Scotland, Iceland, and Greenland have been using this herb for centuries.

The plant grows up to 20 inches (50 cm) tall. The sprouts that are sold in stores are mostly young plants of about 8 inches (20 cm). Oyster leaf is native to salty environments. It has beautiful blossoms that start out red and later turn light blue. By the time it reaches full bloom, this plant looks more like an herb than a vegetable.

It's crunchy, briny, and reminiscent of the taste of oysters, hence the name. Oyster leaf can be used raw in dishes; you only need to rinse the leaves with water. With its intensely herbal flavor it's a real taste sensation. The root is also edible, though you will want to cook it first—frying is best! We also recommend steaming the oyster leaf to better retain its flavor and texture.

Sea kale

As the name may suggest, sea kale (*Crambe maritima*) is related to the cabbage family (Brassicaceae, formerly known as Cruciferae). This plant is native to the coasts of Europe. With its enjoyment of sea breeze, freezing temperatures, and salty soil, sea kale thrives in places where other plants wouldn't survive. It is a large, graceful plant with long stalks, large green leaves, and white flowers. The flowers emit a honeylike odor. The plant itself is fragile and difficult to transport, and it is best eaten just after harvesting. Sea kale has likely been cultivated in France and Great Britain since the eighteenth century. Initially it was only harvested in the wild.

The earthy, salty-nutty flavor reminds us of other types of cabbage. The young sprouts of this plant are especially tasty and can be eaten fresh (they're great in salad!). The long stems can be eaten after they have been cooked for about ten minutes. You can also blanch or stir-fry them. Once the sea kale has become fully grown, it becomes very bitter, akin to rhubarb. The green leaves aren't edible—so don't try!

Sea fennel or rock samphire

Sea fennel (*Crithmum maritimum*) is a plant belonging to the Umbelliferae family (also called Apiaceae) and is native to the coasts of Europe: on the Black Sea, the Mediterranean Sea, the Canary Islands, and the Atlantic Ocean. The plant can grow from 8 to 20 inches (20 to 50 cm) tall and has a woody stem; short, thick leaves (ensuring as little water as possible will evaporate); and yellow-green or white flowers.

Sea fennel is sold by specialized growers and available mainly in the spring. It contains essential oils, which give it a unique, characteristic flavor. Because this plant isn't submerged in seawater permanently, the salty taste is less dominant. Only the fleshy leaves are edible, and they're perfect for salads and other vegetable dishes.

Sea beet

Sea beet (*Beta vulgaris* subsp. *maritima*) is a member of the Amaranthaceae family and is the wild ancestor of common vegetables such as sugar beet, beet (or beetroot), and Swiss chard. The plant grows primarily in Southern Europe, North Africa, and on the Mediterranean coast. It can also be found along the European west coast, all the way up to the southern coasts of Sweden and Norway. The sea beet requires moist, moderately salty soil, and it grows from 12 to 32 inches (30 to 80 cm) tall. It has a slightly nutty and salty flavor. It has rosette-shaped leaves, and the leaves—not the root—are edible. The young leaves have a particularly pleasant taste, while the older leaves tend to be a little bitter. This vegetable, too, lends itself perfectly to steaming.

Ice plant

The common ice plant (*Mesembryanthemum crystallinum*) is a plant from the Aizoaceae family. It's native to the Southern Hemisphere but has been increasingly appearing in the Northern Hemisphere. The ice plant requires temperatures above 70°F (20°C) to grow and therefore thrives in the regions around California, the Mediterranean, and on the Canary Islands. It lives on salty, nutrient-poor soils and has a tremendous ability to draw moisture from the ground. The ice plant is also known by its French name, *Ficoïde glaciale*, and is having a moment in the kitchen. Its leaves and stems are both edible. The leaves are delicious in salads or when cooked just like spinach.

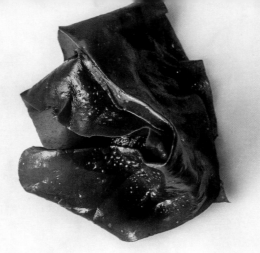

DULSE
(PALMARIA PALMATA)

SEA SPAGHETTI
(HIMANTHALIA ELONGATA)

SAMPHIRE
(SALICORNIA)

DOS & DON'TS

DO: GO PLANT POWER!

Seaweed is sometimes called the protein source of the future. Certain algae species contain protein compounds resembling those found in animal products: a lot of calcium *and* omega-3 fatty acids. After all, fish don't produce omega-3s themselves; they take it in by eating algae. Plant-based cooking is already rich in proteins, iron, and calcium. Adding seaweed to that diet results in a versatile cuisine that isn't just healthy and very delicious but is also sustainable and beneficial to humans, animals, and nature.

DO: EAT WEED, LIVE LONG!

This is The Dutch Weed Burger's maxim and it is well deserved (see page 114). Seaweed: What a magnificent vegetable! And what a luxury that this sumptuousness is there for us to eat. So—go get your 'weed and EAT it!

DO: CHOOSE DULSE & SEA LETTUCE!

Are you still green when it comes to eating seaweed? Do you find it all a bit *too* exciting? Then go find yourself some dulse or sea lettuce. These algae are two of the most accessible and easiest to work with to kick off your seaweed adventures.

DO: MIX & MATCH!

Once your seaweed adventure is well on its way, we recommend including green, brown, and red algae in your diet. Each type offers its own variety of vitamins and minerals, and you want to benefit from all that these seaweeds have to offer.

DO: USE ONLY A LITTLE WATER TO SOAK DRIED SEAWEED

It's a good idea to soak dried algae in just a small amount of water. That way you prevent washing away all the useful vitamins and minerals when you strain the rehydrated seaweed. If needed, you can squeeze out any excess moisture with your hands. The package will normally offer exact soaking instructions. You can assume that tender, succulent algae require less soaking than tough, darker species.

DO: ALWAYS THOROUGHLY RINSE SEAWEED!

When using fresh seaweed, make sure to truly rinse it. You want to be sure that all sand has been washed away and that you won't break a molar on a stray barnacle. Did you pick fresh seaweed directly from the ocean? Then wash it with salt water to prevent it from turning slimy. Preserved seaweed from the organic market must also be thoroughly rinsed, or—better still—soaked. You may opt to pickle your fresh seaweed, which extends its shelf life. At the same time, however, preserving seaweed renders it very salty. After rinsing, gently squeeze out any excess water and pat dry with paper towels or a dishcloth.

DO: FRESH OR DRIED?

What do you do when a recipe calls for dried seaweed and all you have lying around is fresh seaweed? And how do you use dried if a recipe calls for fresh? It's simple, actually. Dried seaweed expands considerably when rehydrated. How much depends on the species, but assume a weight increase of eight to ten times. So if a recipe calls for 0.2 ounce (5 g) dried wakame, that equals 1.6 to 2 ounces (40 to 50 g) fresh wakame. And if it says 1.5 ounces fresh dulse, you know that you can use a bit less than 0.2 ounce dried dulse.

DO: GET IN YOUR KOMBU ZONE!

Kombu is fantastic for making a basic broth. It enriches any soup with a generous dose of umami while seasoning it with natural salt. Apart from that, we recommend adding a piece of kombu when cooking beans. Not only will the kombu enrich the beans with extra nutrients, but the alga will also make them easier to digest.

DO: GO GET YOUR FLAVORS!

Seaweed is rich in umami, which offers many possibilities for further exploration while cooking! There are several herbs and ingredients that boost and complement seaweed's umami flavor. Some of our favorites include nutritional yeast, soy sauce, tahini, miso, mirin, sesame seeds, sesame oil, and tomato.

DON'T: EAT TOO MUCH SEAWEED PER DAY

Seaweed is delicious, and now that you've fallen in love with these ocean greens, you likely can't wait to polish off multiple bowlfuls each day. Understandable, but don't do it! Seaweed is super healthy and it contains all sorts of nutrients. That is precisely the reason you need only a little bit each day. It's a natural source of iodine and omega-3s, for instance. Both are indispensable for human health, but you shouldn't take in too much of them. For the average adult, we recommend 0.2 to 0.35 ounce (5 to 10 g) of dried seaweed per day. As with everything else you eat, consume seaweed in moderation and as part of a diverse and well-balanced diet. Or, in the words of Paracelsus, the famous physician, philosopher, and toxicologist from the 1500s, "All things are poison and nothing is without poison. The dose makes the poison."

DON'T: EAT LOOSE SEAWEED FOUND ON THE BEACH

After having read this book, you may be very enthusiastic to start working with seaweed, perhaps even so enthusiastic that the seaweed washed up on the beach suddenly seems irresistible. However, don't be tempted! You don't know the origins of the algae nor how long it has been lying there. Avoid a bad stomachache from eating rotten seaweed, and never pick algae from contaminated water. Leave the serious picking to the professionals; that way both you and our water habitats, with their natural beauty, can remain in good health.

RECIPES

SPECIAL INGREDIENTS

THERE ARE SEVERAL INGREDIENTS that will really make your seaweed shine. You will see these ingredients, including miso, smoked paprika, Danish smoked salt, Celtic sea salt, nutritional yeast, soy sauce or tamari, sesame seeds, sesame oil, mirin, tahini, garlic, onion, tomato, and *persil de la mer*, throughout the recipes. You can often find them at organic food markets. In this chapter we discuss some of these ingredients that may be new to you.

NUTRITIONAL YEAST

Nutritional yeast is sold at organic food stores as flakes, and it is increasingly found in local groceries. It tastes a little like Parmesan cheese and is an ideal flavor booster for pizza or a bowl of pasta. It's also incredibly healthy. Nutritional yeast is a deactivated yeast that contains many of the B-complex vitamins, such as B1, B2, B3, B6, and folic acid. It helps develop healthy and shiny hair, strong nails, and beautiful skin.

MISO

Miso is a basic ingredient used primarily in soups. It's a salty, fermented bean paste (sometimes with rice, wheat, or rye also added) that functions as a perfect foundation for balancing yin and yang. It's exceptionally popular in "macrobiotic" cooking, a health-based belief system developed from Zen Buddhism. The principle is based on a well-balanced relationship between yin and yang—elementary values used to explain the reaction between all elements of the universe—through food. Miso supports your intestinal flora and also contains protein. In Japan, people often have miso soup for breakfast to ensure a powerful start to the day. You can find all sorts of miso varieties at the organic food market, from very light and mild to intensely dark and strong. Each variety has a different taste and is used in different ways, including in dressings, in soups, and even as a sandwich spread.

SMOKED PAPRIKA

Smoked paprika is indispensable in seaweed cuisine. It lends dishes a deep, meaty flavor. It's almost incomparable to regular paprika powder. The smoked variety is made by smoking a particular kind of peppery paprika over an open fire. It magically converts even the most basic dish into a savory meal. You can find smoked paprika, sometimes also called sweet or Spanish smoked paprika, or *pimentón*, in organic markets and in specialized delicatessen stores like Dean and Deluca.

SALT

The wondrous world of salt: There are so many different kinds of salt with so many different flavors that it's often difficult to decide which one to use. Do you choose coarse or fine sea salt? Table, Celtic, Himalayan, or kosher salt? A thorough examination of high-quality salts may be worthwhile, since a quality salt can immensely influence the taste of a dish. With seaweed we highly recommend using Danish smoked salt and Celtic sea salt, and we advise you to use them for the recipes in this book as much as you can. The texture of salt (coarse or fine) also affects the quality of a dish. Organic food markets usually sell a variety of salts such as *sel gris*, also known as *gros sel marin* (coarse sea salt), and *fleur de sel* (the flower of salts, fine sea salt). Both contain many minerals and have both sweet and bitter characteristics. Out of all salts, Celtic sea salt has the lowest sodium chloride content and is therefore the mildest and "friendliest" salt in the world.

Smoked salt is particularly special: It's processed to have a wonderfully intense, smoky aroma. Try to find a variety that has been smoked over natural fire and is free of toxins. The main ingredient for Danish smoked salt is sea salt from the Dead Sea, which is then smoked in Denmark. The resulting salt is a superior product, but of course any other type of smoked salt also suffices, as long as it's produced in a natural way.

SOY SAUCE AND TAMARI

You can use salt to season your dishes, but soy sauce and tamari are fantastic alternatives to add some saltiness. You only need a tiny bit of either, as they each have a remarkably strong flavor that can enhance the umami in seaweed. Soy sauce and tamari are both fermented soy condiments naturally produced from soybeans, sea salt, mineral water, a special fungal culture, and, in the case of soy sauce, also wheat. The difference between the two is that soy sauce is made from soybeans cooked with wheat before being fermented, while tamari is made from the by-product of miso paste (see page 48). Although wheat is often used to make tamari, it can be made without it—resulting in a gluten-free product. Some soy sauces that are marketed as gluten-free are in fact tamari. It is better to buy organic soy sauce or tamari to ensure that the soybeans have not been genetically modified. Some commercial soy sauce or tamari is chemically produced and contains toxins. You can find these in your local grocery store.

MIRIN

Mirin is a Japanese sweet rice wine. It's made by brewing and fermenting brown, sweet rice with *koji*—a type of fungus that activates the enzymes in the rice. Koji is also used when making sake (a Japanese alcohol), soy sauce, tamari, and miso. Nearly all alcohol in the mirin will evaporate during cooking, while the sweet flavor will remain in the dish. This delicate, sweet flavor pairs exceptionally well with seaweed, miso, and soy sauce or tamari.

PERSIL DE LA MER

Persil de la mer, or sea parsley, is a seasoning made from various types of algae. If you can't find it under this name, look for a seasoning blend of seaweed flakes, usually sea lettuce and dulse, in organic food markets or online at stores specializing in seaweed products. Maine Coast Sea Vegetables, for example, produces Triple Blend Flakes consisting of sea lettuce, dulse, and nori flakes, in their Sea Seasonings line.

Binding & Thickening Agents

Irish moss, carrageenan, and agar are binding agents made from or derived from seaweed. Each has its own character and function, but because of their gelatin-like properties they are great for thickening sauces, crèmes, puddings, frostings, mousses, and cakes.

IRISH MOSS

Irish moss is a red alga species that has been on the rise because of its unique nutritional value. It can absorb a lot of water, which produces a large amount of mucilage (a protein-rich by-product). Mucilage has been attributed various medicinal properties, such as painkilling and anti-inflammatory ones. It regulates the production of mucus in the body, repair damaged mucous membranes, and help move food through the digestive system. Moreover, Irish moss is rich in minerals.

CARRAGEENAN

Carrageenan is a mixture of polysaccharides that is derived from Irish moss and is used mostly in the food industry to solidify or thicken products.

AGAR

Agar is the most well-known of the three. Agar, also called agar-agar, is a whitish, flavorless, and tasteless polysaccharide that is derived from the cell walls of red algae. It is produced in Sri Lanka, Indonesia, and Japan, where it is called *kanten*. In Asian cuisine, agar has been used for centuries as a binding agent. Nowadays you can find it in the West in the form of powders, flakes, and blocks. If you're trying to find an organic version, you'll most likely find it in powder form. Only then can you be certain that the agar is pure and unbleached. If you're already familiar with agar, you'll know it's a challenge to use in cooking. Too much agar powder and you'll end up with a solid food that you can break into pieces; not enough powder and your crème or pudding won't thicken properly. If you want to learn how to work with agar, it's best to experiment in the kitchen before you use it in a recipe. We recommend getting to know how agar works by testing the desired ratio of liquid to powder. Start with less than you expect. Agar powder needs to be heated in the liquid, while stirring, for optimal functionality. Cooling down is essential as well. You may think while stirring that you haven't used enough agar and may be tempted to add more, but test with a smaller quantity and you'll see that once it cools off, the mixture stiffens. If you use agar for a crème, you may find that even after cooling, the result isn't the right consistency. However, if it is too firm, you can puree it in a food processor or with an immersion blender, and you should have your desired consistency in no time!

DIRECTIONS
FOR THE RECIPES

- The measurement for tablespoons is 0.5 fluid ounce (15 ml) and for teaspoons is 0.17 fluid ounce (5 ml). We recommend using standardized measuring spoons, available as a set in every kitchenware store. If a recipe makes use of spoons as a measurement, always assume a level spoon, unless otherwise indicated.

- Ovens differ, even when they are from the same brand. Read the directions for temperature but adjust if necessary for your own oven. It's helpful to always have an oven thermometer on hand.

- All recipes have been tested with 100 percent plant-based and organic ingredients.

- Make sure to taste the food during cooking, where possible. Recipes suggest the flavor, but the best indicator is taste testing throughout the cooking process.

- The shelf-life indicator in recipes is based on working with clean tools, thoroughly covering a product, and storing it in a cool place.

- The authors recommend carefully moderating your intake of seaweed. It should be added as a supplement to your regular diet. For the average person, 0.2 to 0.35 ounce (5 to 10 g) of dried seaweed is a sensible portion. Do eat a varied diet and add seaweed on a regular basis. Don't eat it only occasionally and then consume a large quantity all at once. This has been taken into account for these recipes.

- For information about buying fresh sea vegetables, as well as dried and/or fresh seaweed, see page 167.

- Since dried whole-leaf seaweed may be difficult to measure in cups and spoons, the amounts called for in the recipes often appear in ounces or grams. The table below gives approximate volume conversions for some varieties including dulse, wakame, kombu, or nori, but consider purchasing a kitchen scale to obtain accurate measurements.

  0.7 ounce (20 g) = 1 cup   0.25 ounce (7 g) = ⅓ cup

  0.35 ounce (10 g) = ½ cup   0.2 ounce (5 g) = ¼ cup

TOPPINGS

PESTO FROM THE SEA

MAKES 1 (8-OUNCE) JAR

This green pesto owes its intense and powerful flavor to the use of kombu. It's such an easy recipe that you can experiment with the ingredients as much as you like. The arugula and basil, for instance, can be substituted with any leafy green of your choice. And the pine nuts can be replaced by any other type of nut, such as cashews or hazelnuts. You can even use sunflower seeds! Serve the pesto on crackers, on a grilled vegetable sandwich, or with a bowl of pasta.

1.4 ounces (40 g) fresh kombu
   (or a mix of kombu and wakame)

⅓ cup (50 g) roasted pine nuts

1 small garlic clove, minced

1 cup (25 g) fresh basil
   (including stems)

1 cup (25 g) arugula

2 tablespoons extra virgin olive oil + extra
   for storing

1½ teaspoons lemon juice

1. Thoroughly rinse the kombu with water, making sure all pickling salt has been washed off. Dab dry and cut into pieces.

2. Puree the kombu, nuts, garlic, basil, arugula, olive oil, and lemon juice in a food processor until the mixture forms a smooth paste.

3. Put the pesto in a glass jar, cover with a layer of olive oil, and store in the fridge for up to two weeks. Always serve with a clean spoon to maintain freshness.

SAILOR MUSTARD DRESSING

SERVES 4

This dressing is perfect over your everyday salad or on oven-roasted vegetables. You can also use it as a fresh dip and serve with some radishes and carrots when you're hosting drinks.

1½ teaspoons dried sea lettuce

3 tablespoons coarse mustard (without honey) or plain yellow mustard

3 tablespoons extra virgin olive oil

1 tablespoon apple cider vinegar

1 tablespoon maple syrup

Fine sea salt and freshly ground black pepper, to taste

Mix all the ingredients except the salt and pepper in a bowl, using a fork or a whisk, into a nice smooth sauce. Season with salt and pepper.

EVERYTHING GOES 'WEED MIX

MAKES SEASONING FOR 20 MEALS

Mix, toss, and sprinkle: That is what this recipe comes down to. This mix is easy to make and delicious with all sorts of dishes, from spaghetti Bolognese to cauliflower soup, to pizza and salads or even the good old sandwich. The Danish smoked salt, with its distinct and deep character, gives this mix an extra boost.

1 cup (150 g) raw cashews, peeled and unsalted

¼ cup (15 g) nutritional yeast

2 tablespoons (10 g) persil de la mer (seaweed seasoning; see page 50) or dried sea lettuce flakes

½ teaspoon Danish smoked salt

Combine all ingredients in a food processor and pulse until the mixture becomes fine and crumbly. Store in a closed container in a cool, dark place for up to one month.

CREAMY BABY GREEN PESTO SAUCE

SERVES 2 TO 3

Chlorella goes well with Italian cuisine. It combines excellently with a green salad or simple pasta with sun-dried tomatoes. This pesto is more of a sauce with the microalga blended in! It owes its salty flavor to the alga, which is also rich in omega-3s.

¼ cup (30 g) roasted pine nuts

1 small garlic clove, minced

2 tablespoons extra virgin olive oil

Zest and juice of ½ lemon

1 teaspoon chlorella

½ cup (15 g) fresh basil (including the stems)

2 tablespoons nutritional yeast

3 tablespoons water

Fine sea salt and freshly ground black pepper, to taste

Combine nuts, garlic, oil, lemon zest and juice, chlorella, basil, nutritional yeast, and water in a food processor and pulse into a coarse sauce. Season with salt and pepper.

SAILOR'S BUTTER

SERVES 4

This creamy sailor's butter is delicious on sourdough toast, on a baguette with some soup, or mixed in with some mashed potatoes and vegetables.

1 cup + 1 tablespoon (250 g) vegetable-based margarine, such as Earth Balance Buttery Spread, at room temperature

2 tablespoons (10 g) persil de la mer (seaweed seasoning; see page 50) or dried sea lettuce flakes

Zest from ½ small lemon

1 shallot, minced

1 garlic clove, minced

Fine sea salt and freshly ground black pepper, to taste

1. Combine the margarine, persil de la mer, lemon zest, shallot, and garlic in a blender and puree until combined. Season with a pinch of salt and pepper—taste carefully; this recipe is naturally very salty!

2. Place the butter in a closed container in the fridge for at least one hour to set. Remove from the fridge 10 minutes before serving. The butter will keep for up to 3 weeks in the fridge, or 2 months in the freezer.

NON*FISH SAUCE

MAKES 1 SMALL BOTTLE (ABOUT 8 OUNCES)

A fish sauce made with seaweed instead of fish—that's the secret to this umami-rich condiment. It's perfect with noodles or in an Asian-inspired soup, but it's also a delicious contribution to peanut sauce for satay. By doubling the recipe, you can fill a full-size bottle. But beware: You won't need much of this sauce.

2 cups (500 ml) water

1 ounce (30 g) dried kombu

1 garlic clove, halved

¼ cup + 1 tablespoon (70ml) soy sauce

1 tablespoon vinegar

2 teaspoons lime juice

1½ teaspoons brown sugar

1½ teaspoons freshly ground black pepper

1 teaspoon ginger powder

Pinch of chili powder

1. Bring the water and kombu to a boil in a saucepan. Lower the heat, cover with a lid, and allow to simmer for 20 minutes.

2. Add the rest of the ingredients and again bring to a boil. Lower the heat, cover with a lid, and allow the mixture to simmer for 30 more minutes.

3. After 30 minutes, taste the sauce. If you find it a bit too salty, add water. Turn off the heat and allow it to cool to room temperature. Strain the sauce and store it for up to five days in a sterilized glass bottle that can be sealed airtight.

CHEESY 'WEED SAUCE

SERVES 4

You can serve this creamy sauce with most savory dishes. It makes for a great cheesy sauce over pasta or cauliflower and other vegetables, or a substitute base for a traditional cheese fondue (in that case, make more!). This recipe calls for arrowroot powder, a gluten-free thickener sold in most health food stores. If you need, you can replace it with 1 tablespoon and 1½ teaspoons of potato starch.

0.35 ounce (10 g) dried dulse

Olive oil

4 shallots, minced

1⅔ cups (400 ml) soy, oat, or rice creamer

1 tablespoon + ½ teaspoon mustard

2 tablespoons arrowroot powder

⅓ cup (20 g) nutritional yeast

Fine sea salt and freshly ground black pepper

1. Soak the dried dulse in water for 5 minutes, drain, and thoroughly pat dry.

2. Heat a splash of olive oil in a skillet. Sauté the shallots for 3 minutes over high heat, until golden brown, and then lower the heat. In a bowl, combine the creamer and mustard and mix thoroughly.

3. Add the arrowroot, dulse, and nutritional yeast to the cream mixture and thoroughly combine. While stirring with a whisk, add the mixture to the sautéed shallots in the skillet. Reduce the sauce to thicken it while whisking continuously to prevent any lumps from forming.

4. Season the sauce with salt and pepper. Remember, the dulse itself is quite salty, so this dish doesn't need much extra salt. Serve immediately.

5. Although best eaten fresh, the sauce can be stored for an extra day. If you do so, briefly puree the sauce with an immersion blender before reheating it.

SWEET WAKAME DRESSING

SERVES 2 TO 4

This tahini-wakame sauce makes a wonderful dip for roasted or raw vegetables. It is also a tasty addition to noodles, lentils, or a Moroccan couscous salad. For a richer sauce, use dark tahini instead of white.

0.2 ounce (5 g) dried wakame

¼ cup + 1 tablespoon (75 g) white tahini

¼ cup (60 ml) sesame oil

3 tablespoons soy sauce or tamari

2 tablespoons apple cider vinegar

2 tablespoons maple syrup

Zest from ½ orange

Freshly ground black pepper, to taste

1. Soak the dried wakame according to the instructions on the package, strain, and thoroughly pat dry. Cut into small pieces.

2. In a bowl, combine the tahini, sesame oil, soy sauce, vinegar, maple syrup, and orange zest. Using an immersion blender, pulse into a smooth sauce. Stir in the wakame and season with pepper.

SEA AIOLI

Samphire gives this creamy homemade mayonnaise a nice salty touch. This recipe calls for ground mustard, which we like to grind ourselves with a mortar and pestle. Additionally, you can replace the safflower oil with olive or sunflower oil, but this will give it a different flavor. This aioli is delicious with oven-baked potatoes, french fries, or roasted vegetables.

1 ounce (25 g) samphire

½ cup (120 ml) unsweetened almond or soy milk

2 tablespoons apple cider vinegar

½ teaspoon fine sea salt

1 teaspoon ground mustard

1 teaspoon rice syrup

1 garlic clove, minced

1 cup (250 ml) safflower oil

Freshly ground black pepper, to taste

1. Rinse the samphire and cut it into slightly smaller pieces. Simmer for about 30 seconds in a layer of water in a shallow saucepan. Strain and thoroughly pat dry.

2. Combine the almond milk, vinegar, salt, ground mustard, rice syrup, and garlic in a deep bowl or container and puree into a smooth mixture with an immersion blender. Add the oil to the milk mixture and continue blending. Move the blender around the bowl while processing until you get a nice smooth sauce.

3. Carefully spoon in the samphire. Season with a pinch of pepper.

4. Spoon the sauce into a glass jar with an airtight lid and set aside in the fridge for about 1 hour, allowing the flavors to blend. Properly stored, this aioli will keep for several weeks.

AVOCADO & 'WEED HUMMUS

MAKES 1 FULL (8-OUNCE) JAR

These days, the grilled vegetable and hummus sandwich has become a staple on many menus, and there are thousands of varieties of hummus. What's more, making it is foolproof! Beet hummus, white bean hummus, avocado hummus—anything is possible, and now you can add spicy seaweed to the list!

0.14 ounce (4 g) dried sea lettuce

One 15-ounce (425 g) can chickpeas

1 avocado, sliced

1¼ cup (20 g) fresh cilantro

2 small garlic cloves, minced

3 tablespoons extra virgin olive oil

3 tablespoons lime juice

2 tablespoons coconut milk

2 tablespoons white tahini

1 tablespoon Thai green curry paste
    (or 2 tablespoons for extra spiciness)

Fine sea salt and freshly ground black
    pepper, to taste

1. Soak the sea lettuce according to the instructions on the package. Drain and thoroughly pat dry.

2. Drain the chickpeas. Place them in a food processor along with the rest of the ingredients except for the salt and pepper and pulse until smooth. Carefully taste and season with salt and pepper.

3. Store the hummus in an airtight container in the fridge; it will keep for at least four days.

DULSE SPREAD

This creamy spread is perfect on a cracker but equally delicious on vegetable pasta or toasted sourdough bread with avocado.

½ cup (75 g) raw, unsalted almonds

¾ cup (90 g) sun-dried tomatoes in oil

0.5 ounce (15 g) fresh dulse

½ cup (15 g) fresh basil (including the stems)

½ garlic clove, minced

Pinch of smoked paprika (pimentón)

Juice of ¼ lemon

1. Place the almonds in fresh water, cover, and set aside to soak for a couple of hours or overnight. Strain the almonds. Then drain the sun-dried tomatoes, keeping the oil.

2. Thoroughly rinse the dulse, making sure any salt has been washed off. Dab dry and cut into pieces.

3. In a food processor, pulse together the almonds, tomatoes, dulse, basil, garlic, paprika, lemon juice, and a splash of the reserved oil into a smooth spread. If it's too dry, add a little more of the sun-dried tomato oil. If your food processor isn't powerful enough to smooth the mix, don't worry; a slightly coarse spread tastes great, too!

4. This spread will keep in the fridge for one week in an airtight container.

EARTHY & CREAMY SEA SPREAD

MAKES 1 FULL (8-OUNCE) JAR

This creamy spread with its soothing character and earthy flavor is perfect with crackers and cocktails. You can use it as pastry filling or you can put it on a roasted sourdough sandwich with baby spinach and roasted tomatoes.

1 eggplant, halved

Olive oil for sautéing

1 red onion, finely chopped

9 ounces (250 g) mixed mushrooms, chopped

1 small garlic clove, minced

3.5 ounces (100 g) tempeh (about half a package), finely cubed

2 tablespoons soy sauce or tamari

1½ teaspoon fresh rosemary, minced

2 tablespoons white tahini

1 tablespoon dried sea lettuce or aonori (nori flakes)

Lemon juice, to taste

Pinch of Danish smoked salt (or Celtic sea salt)

Freshly ground black pepper, to taste

1. Preheat the oven to 425°F (220°C) and line a baking sheet with parchment paper.

2. Place the eggplant cut-side down on the baking sheet and cook about 40 minutes, until tender. Remove from the oven and allow to cool until easy to handle. Scrape the pulp from the skin. Chop the pulp finely.

3. Heat a splash of olive oil in a skillet, add the onion, and sauté over medium heat until golden brown, about 5 minutes. Then add the mushrooms and garlic and sauté for another 3 minutes. Spoon in the tempeh cubes and, while stirring, sauté the mixture until the tempeh and mushrooms are crispy and golden brown. Add the eggplant pulp and continue cooking for another minute or so. Douse the mixture with the soy sauce and sprinkle with the rosemary. Continue stirring until the soy sauce is fully absorbed and any excess liquid has evaporated. Remove from the heat.

4. In a food processor or blender, combine the tempeh mixture with the tahini and sea lettuce and pulse until creamy and thoroughly combined.

5. Season with lemon juice, Danish smoked salt, and pepper. This dish is naturally salty and tangy, so be sure to taste it carefully before seasoning.

CREAMY SPIRULINA SAUCE

MAKES 1 (8-OUNCE) JAR

Not only is this creamy dressing ridiculously "green," but it also has a distinct sweet and spicy flavor—a combination of ginger, lime, and cilantro—that makes this condiment perfect for Asian salads and dishes with pasta and coconut. It's also delicious with a salad of sea spaghetti and roasted, grated coconut!

¾ cup (90 g) unsalted pumpkin seeds

1 cup (235 ml) water

1½ cup (25 g) fresh cilantro

1 garlic clove, minced

Zest and juice of 1 lime

2 tablespoons extra virgin olive oil

1.5 cm fresh ginger, minced

1 teaspoon light miso paste, soy sauce, or tamari

1 teaspoon spirulina

Pinch of cayenne pepper

Fine sea salt and freshly ground black pepper, to taste

1. Soak the pumpkin seeds in some water in a covered bowl for 4 to 6 hours, or overnight. Strain them.

2. In a high-powered food processor or blender, thoroughly pulse all ingredients except the salt and pepper into a smooth and creamy sauce. For this recipe it's important to grind the pumpkin seeds as finely as possible. If needed, pause blending halfway through to scrape down the sides of the bowl. Season generously with salt and pepper.

BREN SMITH
THIMBLE ISLAND OCEAN FARM

Before **BREN SMITH** became what he calls "a climate farmer," he spent many years as a fisherman. Through his innovative 3D ocean farm and GreenWave foundation, Smith is making a strong case for a comprehensive model of seaweed cultivation and sustainable development in the US.

You've developed an innovative method of ocean farming; can you describe what a restorative 3D ocean farm looks like?

Imagine a vertical underwater garden, a square of connected ropes floating near the ocean surface with fronds of kelp, *Saccarina latissima* (devil's apron), and other seaweed species growing downward next to scallops, mussels, and oysters buried in the sea floor. Because these farms are vertical and underwater, they have a small footprint and are invisible from the shore. On top of that, they require only a modest investment. Anybody with twenty acres, thirty thousand dollars, and a boat can start one.

This is quite an ambitious plan; would you say you're more of an idealist or a pragmatist?

I'm neither, just as I'm no environmentalist. The reality is, turning our oceans into giant conservation zones won't help without developing technology to mitigate climate change. I'm no foodie, either. My role is to tell the story—both my own history as a farmer and that of the oceans. The key issue for me is how to make a living on a living planet. Real change comes from improving communities as well as the environment, and food is an agent of that change. Seaweed is a zero-input crop, creates income, protects the planet, and yields delicious food.

But seaweed needs further integration into our lives?

Right now our relationship with the sea is strained. We have two separate food systems that are intimately connected but not perceived as so. The way to fix it is by building a bridge between land and sea; our thinking must not stop at the water's edge.

For a non-foodie, you work with chefs quite a lot. What role do they play?

Seaweed captures five times more carbon than land-based plants. It's one of the most exciting food frontiers—and has a low impact on the environment. Fifty years from now seaweed will be the most affordable food on the planet and everyone will be eating it. We're starting now: As climate farmers, we work with chefs to help develop what we think of as a "climate cuisine."

Eating like a fish instead of eating fish?

Exactly. This is why our most successful collaborations aren't necessarily with seafood specialists. Seaweeds are vegetables. People think of them as seafood but they require a different creative eye. With the right mind-set, the possibilities are infinite! Imagine being a chef in 2016 and discovering that there are thousands of vegetable species you've never cooked with. It's like discovering corn, arugula, tomatoes, or lettuce for the first time.

HORS D'OEUVRES

TOFU with SEAWEED CRUNCH

SERVES 4

These crunchy tofu cubes coated with seaweed are delicious, whether as part of a hot meal cooked on the grill or simply served cold for a summer picnic. Try serving them as a snack when hosting a party, perfect with drinks. Cries of delight will fill the air. In any case, this is the kind of recipe you'll have to share.

18 ounces (500 g) firm tofu

1 tablespoon maple syrup

1 tablespoon soy sauce

2 garlic cloves, minced

1 inch (2.5 cm) fresh ginger, finely chopped

1½ teaspoons cumin seeds

Juice of 1 lime

⅓ cup (50 g) flour

⅓ cup (50 g) sesame seeds

2 tablespoons aonori (nori flakes) or crumble or cut one nori sheet into fine flakes

Olive or sunflower oil for panfrying

1. Drain the tofu and gently pat dry with paper towels. Cut the tofu into 1-inch (2 cm) cubes.

2. In a bowl, combine the maple syrup and soy sauce with the garlic, ginger, cumin, and lime juice. Mix in the tofu, making sure that all cubes are covered with the mixture. Cover the bowl and set aside in the fridge to marinate for at least two hours (overnight is even better).

3. The next day, separate the tofu from the excess marinade (you can keep it as a dip). Mix together the flour, sesame seeds, and aonori. Dredge the tofu cubes in the flour mixture, giving them a good coating.

4. Heat a few tablespoons of oil in a skillet. Panfry the tofu until golden brown on all sides, flipping occasionally. Spoon the cubes onto a plate lined with paper towels and allow to drain.

5. Serve the crunchy tofu cubes hot or cold. Add some Non*Fish Sauce (page 59) dip if you like.

FAST & SIMPLE SAMPHIRE or SEA ASTER

SERVES 4

The mild, briny tang of samphire and sea aster are easy on the palate for people with diverging flavor preferences. Aside from that, they contain many useful minerals, which makes this tart recipe a healthy but also tasty addition to a lot of dishes. It's great in salad; fresh, light pasta; or served as a side dish with the Everything Goes 'Weed Mix (page 56).

Olive oil for sautéing

1 small shallot, diced

1 small garlic clove, minced

10.5 ounces (300 g) fresh samphire or sea aster, rinsed and dabbed dry

Juice of ¼ lemon

1 tablespoon water

1 tablespoon capers (optional)

Freshly ground black pepper, to taste

1. Heat a splash of olive oil in a skillet. Sauté the shallot until golden brown, about 3 minutes. Add the garlic and sauté for another minute. Add the samphire or sea aster and stir-fry for about 2 minutes. Douse with the lemon juice and the water. Allow to simmer for about 3 minutes over medium heat.

2. Prepare the capers, if using. Heat a small splash of olive oil in a skillet. Scoop the capers from the jar and dab dry with a paper towel. Add the capers to the hot oil (watch out for splashing) and fry until crispy.

3. Serve the steamed samphire or sea aster with the fried capers and season with black pepper.

WARM DULSE & FAVA BEAN SALAD

SERVES 4

Fava beans fresh from the pod, tender dulse, and tangy lemon along with sultry sage and thyme seasoning make this the *perfect* condiment for a spring or summer dish. Serve with a simple pasta or green salad, or double the measurements and serve with a hearty stew. The Everything Goes 'Weed Mix (page 56) is great for finishing this dish.

1.75 ounces (50 g) fresh dulse or 0.2 ounces (5 g) dried dulse

Olive oil for sautéing

2 shallots, minced

1 garlic clove, minced

10.5 ounces (300 g) fresh fava beans, shelled

1 tablespoon fresh thyme

1 cup (235 ml) soy, almond, or rice creamer

Juice of ¼ lemon

3 tablespoons capers (optional)

12 fresh sage leaves

Fine sea salt and freshly ground pepper, to taste

1. Carefully wash the dulse, making sure to rinse off any brine residue. Pat dry and cut into pieces.

2. Heat a splash of olive oil in a skillet, add the shallots, and sauté over medium heat until golden brown, about 3 minutes. Add the garlic and sauté for another minute. Then add the fava beans, dulse, and thyme and sauté for about 2 minutes, stirring occasionally. Douse with the soy creamer and lemon juice. Simmer for 5 more minutes over low heat, with the pan partially covered.

3. In the meantime, spoon the capers from the jar, if using, and dab dry with a paper towel. Heat a splash of olive oil in a skillet. Add the capers (be careful, the oil can splatter) then add the sage and fry until the capers are crispy and the sage leaves are crunchy.

4. Serve the fava bean–dulse mixture with the sage and capers and season with a pinch of salt and pepper.

TEMPEH SEAWEED SNACK

SERVES 2 TO 4

This snack is just perfect served with drinks. Add some cocktail skewers and success is guaranteed! They're simple, easy to make, and deliciously salty and crunchy. Also terrific in a hearty soup or stew. The smoked paprika is essential to elevate this recipe's flavor.

One 8-ounce (225 g) package tempeh

Olive oil for sautéing

1 tablespoon soy sauce

1½ teaspoons aonori (nori flakes) or crumble or cut ½ nori sheet

Pinch of smoked paprika (pimentón)

1. Cut the tempeh into small cubes.

2. Heat a splash of olive oil in a skillet and sauté the tempeh until crunchy and golden brown all around, 6 to 7 minutes.

3. Add the soy sauce (watch out for splattering oil) and thoroughly mix so the cubes can absorb the soy sauce.

4. Sprinkle with the nori flakes and smoked paprika. Continue cooking for another minute, until the seasoning and tempeh are nicely combined.

EGGPLANT CAVIAR

SERVES 4

This fresh, tangy, and creamy spread is ideal on a cracker, on a baguette with arugula, or as a side dish to a wonderful Italian-inspired meal.

2 medium eggplants, halved

Olive oil for sautéing

1 large onion, finely chopped

1 garlic clove, minced

1 tablespoon lemon juice

1 tablespoon tomato paste

1½ teaspoons balsamic vinegar

1½ teaspoons persil de la mer (seaweed seasoning; see page 50)

Pinch of Danish smoked salt (or Celtic sea salt)

Freshly ground black pepper, to taste

1. Preheat the oven to 425°F (220°C) and line a baking sheet with parchment paper.

2. Halve the eggplants lengthwise and place cut-side down on the baking sheet. Cook for about 40 minutes, until tender. Remove from the oven and allow to cool until easy to handle. Scrape the pulp from the skin. Finely chop the pulp.

3. Heat a splash of olive oil in a skillet, add the onion, and sauté over medium heat until golden brown, about 5 minutes. Add the eggplant pulp and the garlic and stir-fry until all oil has been absorbed.

4. Add the lemon juice, tomato paste, vinegar, and persil de la mer. Carefully stir-fry the mixture for another minute or two.

5. Season with Danish smoked salt and pepper. The "caviar" should have a fresh, tangy, and lightly salted flavor.

SHIP AHOY PICCALILLI

MAKES 2 LARGE (1 QUART) JARS, OR SEVERAL SMALLER ONES

Sea spaghetti preserved with cauliflower, bell pepper, carrot, onion, and all sorts of exciting herbs forms the base of this quite extraordinary piccalilli. Take ample time when preparing this condiment, because the vegetables need to rest in a layer of salt for about eight hours. This recipe can be canned and stored for later though be sure to brush up on best practices for canning to avoid bacteria. You can serve this tangy, briny sauce to go with fries or a bean stew.

0.7 ounce (20 g) dried sea spaghetti

1 carrot, chopped

1 red bell pepper, diced

1 red onion, finely chopped

1 small cauliflower, cut into florets

½ fennel bulb, chopped

4 teaspoons fine sea salt

Olive oil for frying spices

2 tablespoons ground mustard

2 tablespoons ground turmeric

1 tablespoon mustard seed

2 bay leaves

1 heaping teaspoon ground ginger

1 teaspoon black peppercorns

Pinch of smoked paprika (pimentón)

3⅓ cups (800 ml) apple cider vinegar

2 cups (500 ml) water

3 tablespoons cane sugar (or maple syrup)

¼ cup + 1 tablespoon flour or ¼ cup arrowroot powder (to be gluten-free)

2 large or several smaller sterilized glass jars for canning*

1. Sterilize the empty glass jars after washing them thoroughly: Place them in a boiling-water canner with a rack on the bottom, filling with warm water to 1 inch above the tops of the jars. Bring the water to a boil and boil 10 minutes. Reduce the heat and keep the jars in the hot water until it is time to fill them. Wash and preheat the lids as well.

2. Cook the sea spaghetti following the instructions on the package. Strain and cut the strings into slightly shorter pieces.

3. Put the sea spaghetti, carrot, red pepper, red onion, cauliflower, and fennel in a bowl and mix in the salt. Cover with a dishcloth and let stand overnight (or about 8 hours) to draw out the liquid. Afterward, pour out any excess liquid.

4. Heat a splash of olive oil in a large pan over medium-low heat and briefly fry all the spices for a nice flavor boost. Then add the vinegar, water, and vegetables. Bring to a boil and cook until the vegetables are tender, 10 to 15 minutes.

5. Drain the vegetables, reserving the liquid. Pour the liquid back into the pan and place over a low flame. Add the sugar and let it dissolve (or, if using maple syrup, stir it in thoroughly). Bring back to a boil.

6. Mix the flour (or arrowroot powder) in a small bowl with 2 to 3 tablespoons of the cooking liquid to make a thickener (mixing until there are no lumps). Lower the heat. Fold the slurry into the pan and keep stirring while the mixture thickens, about 5 minutes. Once the sauce has a nice consistency, add the vegetables and seaweed and stir thoroughly. Remove from the heat and allow to cool completely.

7. Spoon the piccalilli into the jars and carefully close each one with a lid. Store them in the fridge. The piccalilli will keep for four to six weeks.

WAKAME & DATE TAPENADE

MAKES 1 (12-OUNCE) JAR

A tapenade made from wakame, dates, red onion, and pecan may sound a little unusual, but don't take our word—try it yourself—it's delicious. The sweetness of the onion and dates perfectly complements the saltiness of the wakame, while the acidity of the balsamic vinegar adds an extra layer of complexity to boost the flavor. Serve as a filling for grilled portobello mushrooms, as a condiment on a crunchy tempeh burger, or with a roasted squash dish.

0.14 ounce (4 g) dried wakame

Olive oil for sautéing

2 red onions, sliced into thin rings

1 small garlic clove, minced

6 dates, pitted and chopped

1½ teaspoon dried rosemary

1½ teaspoon balsamic vinegar

1 teaspoon maple syrup

Coarse sea salt and freshly ground pepper, to taste

⅓ cup (35 g) unsalted pecans

1. Soak the wakame, following the instructions on the package. Strain and pat dry.

2. Heat a splash of olive oil in a skillet, add the onions, and sauté over medium heat until golden brown, about 5 minutes. Add the garlic and sauté for another 2 minutes. Add the wakame, dates, and rosemary and cook for another 5 minutes. Sprinkle with the vinegar, thoroughly stir, and douse with the maple syrup. Allow the vinegar and maple syrup to caramelize. Season with salt and pepper and remove from the heat.

3. Toast the pecans in a separate pan over low heat until crunchy. Be careful not to burn them; sometimes they cook very quickly.

4. Spoon the pecans over the tapenade and serve.

MAINE COAST SEA VEGETABLES

In 1971 seaweed pioneers Linnette and Shep Erhart began selling *Alaria* fronds they had dried over their woodstove to friends and neighbors. Today, with their daughter Seraphina at the helm, **MAINE COAST SEA VEGETABLES** still thrives on the same spirit of curiosity and sustainability.

What do you make of the current wave of seaweed enthusiasm?

SHEP: I feel it's part of a rising consciousness about health, of our own bodies and of the planet as a whole. And I think it's very welcome because the SAD diet, the Standard American Diet—you've heard of it, I'm sure—is in dire need of improvement. No wonder people are looking for alternatives. Of course there are many nutrient-dense superfoods, but today's environmentally conscious consumers are no longer willing to settle for stuff shipped in from the far corners of the world. They are asking themselves: *What is available locally?* The answer for so many people: seaweed.

How has the awareness of seaweed changed since you went into business?

We've seen a lot of changes in the forty-five years since we sold our first seaweed. Seaweed consciousness has been growing globally and these past few years there's been more local food awareness. Our customers are extremely curious; they love learning about these plants, their habitat, and where their food is coming from. In the last ten years we've been overwhelmed with requests by amateur sea farmers and harvesters who want to learn from or work with us. We hope they will approach the sea as we did, thinking of sustainability from the start and not taking too much right out of the gate.

What about the concern of eating too much of it?

SERAPHINA: With my mother, Linnette, eating seaweed when she was pregnant with me, I might be one of the few American children born with many necessary digestive enzymes. Most of us in the West lack these, though. Seaweed is a superfood in the sense that a little goes a long way. It can taste really good and your body craves the nutrients and minerals, but if you eat too much—and I've seen this with new staff who get really excited—you can get a stomachache. We always tell people that variety is key. Just like with any other vegetable. Alternate between green, brown, and red algae, for instance. Each has its own specific benefits. Many people are seeking seaweed for a particular piece of the nutritional puzzle, but its benefit is much more than the sum of its nutrient values.

How has contemporary seaweed farming changed the way we eat seaweed?

Although we haven't tested them, we're beginning to see some essential differences between fresh and farmed seaweeds. For instance, with farmed seaweeds the plants often seem more tender and mild. We're excited for what these changing flavors might mean for chefs and home cooks!

SOUP

JAPANESE DASHI & NORI CRISPS

SERVES 4

Dashi, a classic Japanese broth, is known for its deep umami flavor. Plant-based ingredients like seaweed, miso, and soy sauce possess that necessary intensity; real, traditional dashi is therefore made with kombu. Naturally, this book had to include at least one deliciously umami-rich dashi recipe. The broth is fantastic as a light meal or snack, delicious both hot and cold. Serve with some wakame and you have a perfect late-afternoon energy booster. You can also turn this broth into a meal by serving it with the carrots and the shiitake mushrooms you used when making it or by adding other types of vegetables, wakame, or pasta. It can also be served with braised leeks, onions, grilled tofu, roasted cashew nuts, and sesame seeds. You can also use the broth as a base for sauces and dressings. The nori crisps are a little bonus, on the house.

BROTH
Sesame oil for sautéing

2 onions, coarsely chopped

1½ inch (4 cm) fresh ginger, minced

3 garlic cloves, minced

2 carrots, cut into 1 cm cubes

8 cups (2 L) water

1.75 ounces (50 g) dried shiitake mushrooms

0.7 ounces (20 g) dried kombu strips

¼ cup (70 g) dark miso paste

3 tablespoons mirin

1 tablespoon soy sauce or tamari

NORI CRISPS
2 sheets dried nori

Sesame oil for toasting nori

White and/or black sesame seeds

Coarse sea salt

1. Preheat the oven to 320°F (160°C) and line a baking sheet with parchment paper.

2. Heat a layer of sesame oil in a stockpot and sauté the onions until golden brown, about 5 minutes. Add the ginger, garlic, and carrots. Sauté for about 5 minutes. Add 4 cups of water and the dried mushrooms and cook for about 30 minutes over low heat. Allow to simmer with the pot partially covered. After 30 minutes, add the kombu and cook over low heat for another 30 minutes.

3. While the broth is simmering, make the nori crisps. Shred or cut the nori sheets into pieces, brush with a little sesame oil, and sprinkle with the sesame seeds and some sea salt. Toast the sheets in the oven for about 5 minutes. Store in an airtight container or serve immediately in the broth.

4. When the broth is done, remove the kombu. Mix several tablespoons of hot soup in with the miso to make a paste and then add it to the soup. Add the mirin and a dash of soy sauce to the broth. (Add the soy sauce in small amounts while tasting carefully!) Thoroughly stir and let rest, allowing the soup to cool and the flavors to blend.

5. Strain the soup through a cheesecloth-lined sieve. Also strain the vegetables and herbs in cheesecloth and squeeze well. Make sure to collect all of the liquid. Now you have dashi!

CHEESY CAULIFLOWER SOUP
with HAZELNUT & SAMPHIRE

SERVES 4

This soup has a smooth and creamy texture. The hazelnuts add a nutty, slightly bitter touch, while the samphire gives it a hint of salt. Serve with a piece of toasted sourdough baguette dipped in Sailor's Butter (page 58) and a nice glass of red wine.

⅔ cup (90 g) unsalted hazelnuts

Olive oil for sautéing

1 large onion, finely chopped

1 cauliflower, chopped into small florets

2 garlic cloves, minced

5¼ cup (1.25 L) water

2 cubes vegetable bouillon

2.6 ounces (75 g) samphire, rinsed, any tough stems removed

Freshly ground black pepper, to taste

2 tablespoons nutritional yeast

Juice of ½ lemon

Fine sea salt, to taste

EXTRAS

Nutritional yeast, optional

Everything Goes 'Weed Mix (page 56), optional

1. Preheat the oven to 350°F (175°C). Line a baking sheet with parchment paper and spread the hazelnuts on top. Roast the hazelnuts in the oven about 12 minutes, until golden brown. Remove from the oven and let rest.

2. Meanwhile, prepare the soup. Heat a generous splash of olive oil in a stockpot. Sauté the onion until golden brown, about 5 minutes. Add the cauliflower florets and garlic and cook for another 5 minutes. Pour in the water and stir in the bouillon. Bring the water to a boil and stir until the bouillon is completely dissolved. Partially cover the pan with the lid and let the soup simmer over low heat for about 20 minutes.

3. Prepare the samphire: Heat a splash of olive oil in a skillet. Sauté the samphire for 5 minutes or so. Remove from the heat and season with pepper.

4. Pour the broth into a blender or food processor along with the hazelnuts, nutritional yeast, and lemon juice (beware of splashes; the soup is hot). Pulse the soup until smooth and creamy. Alternatively, you can add the nutritional yeast and lemon juice to the stockpot and blend with an immersion blender.

5. Carefully taste and season with a pinch of sea salt. Serve the soup with the fried samphire and some extra nutritional yeast or—if you have it on hand—some of the Everything Goes 'Weed Mix.

SEA CHOWDER & SHIITAKE BACON

SERVES 4

Sea-based chowders are a New England staple. Now, vegans can enjoy this slightly sweet, perfectly salty treat, too. This chowder uses sweet and summery corn, and its creaminess comes from potatoes. The saltiness is courtesy of a true king of the sea: kombu. The shiitake bacon with its intense, savory flavor tops it all off.

SEA CHOWDER
Olive oil for sautéing

1 large white onion, finely chopped

3 garlic cloves, minced

1 small red chile pepper, seeded and minced

3½ cups (500 g) fresh corn kernels (from about 2 ears)

3 medium potatoes, peeled and cubed

1 teaspoon thyme

2 bay leaves

5¼ cups (1.25 L) water

0.25 ounce (7 g) dried kombu

Fine sea salt and freshly ground pepper

SHIITAKE BACON
Olive oil for sautéing

7 ounces (200 g) shiitake mushrooms, wiped clean and thinly sliced

2 tablespoons soy sauce

1 heaping teaspoon smoked paprika (pimentón)

2 to 3 teaspoons maple syrup

EXTRAS
Fresh herbs, to taste (dill weed, parsley, rosemary, cilantro, etc.)

1. **To make the sea chowder,** heat a splash of olive oil in a stockpot. Sauté the onion until golden brown, about 5 minutes. Add the garlic and the chili pepper and cook for 3 more minutes. Add the corn, potatoes, thyme, and bay leaves. Stir for a couple of minutes. Add the water and kombu. Bring to a boil and lower to a simmer for 20 minutes, the pot partially covered with a lid.

2. **To make the shiitake bacon,** heat a splash of oil in a skillet and sauté the shiitakes until golden brown all around. Douse with the soy sauce and stir, allowing the shiitake mushrooms to fully absorb the liquid. Sprinkle with smoked paprika and the maple syrup. Stir thoroughly and allow the mushrooms to caramelize (this process goes quickly, so be careful). Remove from the heat.

3. Remove the kombu from the soup. Cook for another 5 minutes, uncovered. Remove from the heat. Scoop out the bay leaves. Puree the soup with a stand or immersion blender until completely smooth. Season with sea salt and pepper.

4. Serve in bowls together with the shiitake bacon and garnish with fresh herbs.

POWER-PACKED GREEN PEA SOUP with SEA LETTUCE

SERVES 4

Besides being deliciously fresh and creamy, this soup is also terrifically healthy. The humble green pea is a powerhouse, and this combination of peas, fiber- and protein-rich sea lettuce, and avocado—the secret ingredient that lends this soup its creaminess—will make you feel big and strong.

2 tablespoons coconut oil

2 medium white onions, finely chopped

2 cups (300 g) green peas, frozen or fresh

2 garlic cloves, minced

2½ cups (600 ml) water

1 cube vegetable bouillon

One 13.5-ounce (400 ml) can coconut milk

0.25 ounce (7 g) dried sea lettuce

2 avocados, sliced

3 tablespoons freshly chopped mint

Juice of 1 lime

Sea salt and freshly ground black pepper, to taste

EXTRAS

Fresh mint leaves, for garnish

Coconut milk

Sea lettuce flakes

1. Heat the coconut oil in a stockpot. Sauté the onions until golden brown, about 5 minutes. Add the peas and garlic and sauté for 3 minutes. Add the water, bouillon cube, and coconut milk. Stir continuously until the bouillon cube has dissolved, and bring to a boil. Continue simmering for another 10 minutes over low heat, the pot partially covered with a lid.

2. Add the sea lettuce to the soup and remove from the heat. Add the avocados, mint, and lime juice. Puree the soup with an immersion blender until smooth. Season with salt and pepper. Note: The soup already contains bouillon and seaweed, so taste carefully before you add any salt!

3. Serve with fresh mint leaves, a splash of coconut milk, and a few flakes of sea lettuce to garnish.

THE FABULOUS FISHLESS SOUP

SERVES 4

This recipe is inspired by a traditional fish soup from the French region of Brittany. Seaweed isn't just a wonderful provider of umami; it's also bursting with omega-3s. These essential fatty acids are one of the main reasons for including fish in our diet. Did you know that when replacing fish with seaweed in this recipe—essentially making a fish soup without fish—you're receiving as much as you need of this must-have nutrient? The crispy fried tofu, oyster mushrooms, and oyster leaves, doused with white wine, make for a sumptuous soup. As an alternative to oyster leaves, we recommend sea aster (6 leaves). Serve with toasted sourdough bread and Sailor's Butter (page 58).

2 beefsteak tomatoes

Olive oil for sautéing

2 onions, finely chopped

2 carrots, cubed

1 fennel bulb, julienned

2 garlic cloves, minced

¾ cup (150 g) dried red lentils

1 tablespoon lemon zest

5¼ cups (1.25 L) water

0.3 ounce (8 g) dried kombu

1 tablespoon dried thyme

7 ounces (200 g) smoked tofu, cubed

5 ounces (150 g) oyster mushrooms, sliced

5 oyster leaves, wiped clean

Splash of white wine

Fine sea salt and freshly ground black pepper, to taste

EXTRAS

Coconut milk

Fresh dill weed

Pomegranate seeds

1. Peel the tomatoes: Using a knife, carve an X in the bottoms of the tomatoes and dip them in boiling water for 10 to 15 seconds. Briefly let them rest, then blanch them in cold water (or an ice bath). Now you can easily peel off the skin using your fingers or a small paring knife. Chop the tomato flesh into cubes.

2. Heat a splash of olive oil in a stockpot. Sauté the onions until golden brown, about 5 minutes. Then add the tomatoes, carrots, fennel, garlic, red lentils, and lemon zest. Add the water, kombu, and thyme. Bring to a boil, then lower the heat and let simmer for about 20 minutes.

3. Meanwhile, heat a splash of oil in a skillet. Sauté the tofu and oyster mushrooms until golden brown. Add the oyster leaves and stir for about 3 minutes Add a splash of white wine. Over medium heat, allow the wine to be absorbed and the extra liquid to evaporate while stirring continuously. Season with salt and pepper. Note: The oyster leaves are themselves salty!

4. Remove the kombu from the soup after 20 minutes. Turn off the heat. Puree the soup with an immersion blender until smooth.

5. Serve the soup topped with the tofu and a dash of coconut milk. Garnish with some fresh dill and pomegranate seeds.

KATHY ANN MILLER

KATHY ANN MILLER is curator of algae at UC Berkeley's University Herbarium. She caught the seaweed bug during an undergrad summer course in 1976 and has been collecting and researching native Californian seaweed ever since.

You've devoted over forty years to researching seaweed; what first interested you about them?

They're the foundation species, the engineers that created their environment! I was immediately fascinated by their form, their otherworldliness; how seaweeds occupy space, how they reproduce; the efficiency with which they capture carbon. Their sheer dominance on earth is incredible. At first they seem static but in fact they move— they throw their spores out into the ocean. They're intriguing, complex, and still mysterious.

As a phycologist, what are your thoughts on the growing interest in seaweed as a food source?

It's wonderful to see more and more people becoming aware of the magnificent creatures growing in our oceans. Eating seaweed also ties in to a broader tendency toward eating lower on the food web. We're moving away from eating animals and instead embracing plants.

Can you tell us about the herbarium?

Our herbarium started in 1895. I think the importance of comprehensive collections like ours is that they are treasure houses of history. These are actual specimens, some of them over a hundred years old, containing a wealth of information only now being revealed to us through modern technology such as DNA sequencing. We're holding information about history— biodiversity and evolution—that will be immensely valuable to the future. We know, for instance, that when explorers sailed north on the Pacific, whole communities of plants and animals hitchhiked on their ships to the New World.

So seaweed as stowaways and storytellers—what can we learn from them?

Flexibility! Land creatures are rigid; they need a skeleton to support themselves in the air. In the ocean, resistance is futile. You need to be flexible. Seaweeds can be eaten, then they grow back. They can be battered, then they adapt. Because of this, they've been around for millions of years, through multiple ice ages and continental shifts. Yet many of them still have forms similar to those they evolved into long ago. They also live alongside other species like animals. They are role models of resilience, adaptability, and collaboration.

When I first started looking at seaweeds and glimpsed this other world, I felt so terrestrial! We tend to be so involved with our own species. It's good to step away from our own perception of the world and see it from another species' perspective. I often try to think like a weed, to see the world in a different way.

SALADS

BEET & 'WEED SALAD

SERVES 4

This sweet and sour salad not only makes for a colorful addition to your table, but it's also wonderfully fresh, crunchy, and spicy. It's perfect for lunch in spring or late summer, or as a side dish for a more substantial meal.

1.75 ounces (50 g) fresh sea lettuce, rinsed and dabbed dry

3 tablespoons olive oil

2 teaspoons balsamic vinegar

2 carrots, peeled

1 large beet, peeled

1 fennel bulb, outer layer removed

1 chile pepper, seeded and finely chopped

Handful of fresh cilantro, finely chopped

DRESSING

1 tablespoon prepared horseradish

Juice of ½ lime

1 teaspoon maple syrup

1 teaspoon extra virgin olive oil

Fine sea salt and freshly ground black pepper, to taste

1. Preheat the oven to 320°F (160°C). Line a baking sheet with parchment paper. Tear the sea lettuce into pieces and sprinkle them with olive oil and balsamic vinegar. Bake for about 15 minutes, until crispy. Keep a close eye on the leaves, as they can burn suddenly.

2. Meanwhile, prepare the salad. Using a mandoline or julienne peeler, julienne the carrots, beet, and fennel. Combine the vegetables in a bowl and add the chile pepper and cilantro.

3. To make the dressing, place the horseradish in a small bowl and stir in the lime juice and maple syrup. While stirring, add the olive oil until you have a nice smooth mixture.

4. Stir the dressing into the vegetables, season with some salt and pepper, and serve with the warm, crispy sea lettuce.

WILD & LUSCIOUS PUMPKIN SALAD with SEA SPAGHETTI

SERVES 4

Say *goodbye* to winter and *hello* to spring with oven-roasted winter squash, caramelized red onion, and lemon slices with a creamy sea spaghetti and fresh, zesty pesto combined in a beautiful salad. Serve with warm slices of sourdough baguette and a glass of good wine.

SALAD

1 kabocha or Hokkaido squash, seeds removed, sliced or diced

1 large red onion, sliced

½ lemon, thinly sliced

1 tablespoon fresh rosemary

Olive oil for roasting vegetables

Coarse sea salt and freshly ground black pepper, to taste

1 ounce (30 g) dried sea spaghetti

3¾ cups (75 g) arugula, rinsed and dabbed dry

1 teaspoon nutritional yeast, optional

PESTO

¼ cup (30 g) pumpkin seeds

3 cups (75 g) arugula, rinsed and dabbed dry

1⅓ cups (35 g) fresh basil

1 tablespoon nutritional yeast

2 to 3 tablespoons olive oil

Lemon juice, to taste

Fine sea salt, to taste

1. Preheat the oven to 400°F (200°C) and line a baking sheet with parchment paper. Combine the squash with the onion, lemon, and rosemary and sprinkle generously with olive oil. Season with salt and pepper. Bake for about 25 minutes (depending on the size of the pumpkin pieces), until brown and crispy.

2. Prepare the sea spaghetti according to the directions on the package.

3. To make the pesto, toast the pumpkin seeds in a skillet until golden brown. In a food processor, combine the arugula with the pumpkin seeds, basil, and nutritional yeast. Pulse until thoroughly combined. Pour in the olive oil and blend. The pesto will be chunky. Season with lemon juice and a pinch of salt.

4. When the squash has become crispy and tender, remove from the oven and allow to cool. Combine with the sea spaghetti.

5. Place the arugula in a large bowl and arrange the pumpkin and sea spaghetti on top. Garnish with some nutritional yeast, if desired, and serve with arugula pesto.

CUT THE CRAB SALAD

A luxurious and creamy potato salad, enriched with sweet and briny dulse, this is perfect for a barbecue or to complement a summer picnic. Serve the potato salad with fresh greens. We like to include radish leaves, extra radish pieces, sliced avocado, and some juice and zest of a lemon.

0.1 ounce (3 g) dried dulse or 1 ounce (30 g) fresh dulse

3 large potatoes, peeled and grated

Olive oil for baking

Fine sea salt and freshly ground black pepper, to taste

3 tablespoons vegan mayonnaise

1½ teaspoons mirin

Pinch of wasabi paste

3 radishes, quartered

1 chile pepper, seeded and finely chopped

1 heaping teaspoon capers, drained and halved

1 heaping teaspoon dried dill or 2 teaspoons finely chopped fresh dill

1. If you're using fresh dulse, rinse thoroughly to get rid of the brine. Pat dry and chop finely.

2. Preheat the oven to 400°F (200°C). Line a baking sheet with parchment paper. Carefully dab dry the grated potato. Coat the potato with some olive oil and arrange on the baking sheet. Bake for 20 to 25 minutes, until golden brown. Season with salt and pepper. Allow to cool for 5 minutes.

3. In a large bowl, combine mayonnaise, mirin, and wasabi and mix well. Stir in the potatoes, dulse, radishes, chile, capers, and dill.

THE HEALTHY & HAPPY BOWL

SERVES 4

This is actually a Buddha Bowl: a generously filled bowl with lots of roasted or raw vegetables, beans, grains, nuts, and seeds, often topped off with lettuce, brussels sprouts, avocado, and a rich dressing. When overflowing, this salad resembles Buddha's round belly, hence the name. What's great about this salad is that you can add anything you like, so indulge and create your own Healthy & Happy Bowl!

3 large carrots, sliced

3 sweet potatoes, peeled and cubed

5 ounces (150 g) brussels sprouts, ends trimmed, halved

One 15-ounce (425 g) can of chickpeas, drained and dried

Olive oil for roasting vegetables

Coarse sea salt and freshly ground black pepper, to taste

1 heaping teaspoon smoked paprika (pimentón)

1 teaspoon thyme

1 teaspoon ground turmeric

1 teaspoon za'atar (optional)

WAKAME

2 tablespoons sesame oil

2 tablespoons soy sauce

2 teaspoons mirin

3 ounces (80 g) fresh wakame, rinsed, dried, and chopped, or 0.3 ounce (8 g) dried wakame, soaked

2 tablespoons sesame seeds

EXTRAS

3 cups (75 g) arugula, rinsed and dried

1 cup (150 g) quinoa or buckwheat groats

2 small avocados, sliced

Juice of 1 lemon

2 tablespoons Everything Goes 'Weed Mix (page 56)

1. Preheat the oven to 400°F (200°C). Line a baking sheet with parchment paper. In a bowl, combine the carrots, sweet potatoes, sprouts, and chickpeas. Stir in enough olive oil to lightly coat all the vegetables. Season with salt, pepper, and the rest of the herbs and spices. Arrange on the baking sheet and roast for about 30 minutes, until the vegetables are golden brown and crispy.

2. Meanwhile, combine the sesame oil, soy sauce, and mirin. Stir well and then add the wakame. Mix thoroughly. Sprinkle with sesame seeds. Set aside for at least 10 minutes so the various flavors can blend. Stir briefly before serving.

3. Meanwhile, prepare the quinoa or buckwheat according to instructions on the packaging. Allow to rest in the pot, covered with a lid until ready to serve.

4. Once the vegetables and chickpeas are tender, remove from the oven and allow them to cool for 5 minutes.

5. Arrange a handful of arugula in each of four bowls. Add the roasted vegetables, chickpeas, and the quinoa or buckwheat. Garnish with some avocado slices and a scoop of wakame. Sprinkle with lemon juice and finish off with some Everything Goes 'Weed Mix.

WINTER 'WEED & KALE SALAD with CARAMELIZED APPLES

SERVES 4

Salads are delicious not only in the spring and summer, but they can also serve as a valuable source of energy in the winter, as this salad proves. Raw, massaged kale combined with protein-rich buckwheat and a wintery apple provide extra vitamin C; finished with parsley, watercress, and seaweed flakes, this is an essential salad for winter.

1 tablespoon olive oil

1 onion, finely diced

1 garlic clove, finely diced

9 ounces (250 g) kale, rinsed and sliced

1 ounce (30 g) fresh dulse or 0.1 ounce (3 g) dried dulse, soaked and dried

2 tablespoons extra virgin olive oil

Zest and juice of ½ lemon

¾ cup (125 g) buckwheat groats

Coarse sea salt and freshly ground black pepper, to taste

CARAMELIZED APPLE

1 small apple

Olive oil or coconut oil for sautéing

SEAWEED TOPPING

½ cup (70 g) shelled pistachios, unsalted (or almonds)

Pinch of Danish smoked salt

1 tablespoon dulse flakes

1 tablespoon sea lettuce flakes

1 tablespoon nutritional yeast

EXTRAS

Handful of fresh parsley, finely chopped

Watercress, for garnish

1. Heat 1 tablespoon of olive oil in a skillet and sauté the onion for 5 minutes, until golden brown. Add the garlic and sauté for 1 minute. Remove from the heat.

2. Put the kale in a large bowl and massage the leaves for 5 to 10 minutes until they are tender. Mix with the extra virgin olive oil and lemon zest and juice to taste.

3. Thoroughly rinse the dulse and make sure to remove all the brine. Pat dry and chop finely.

4. Following directions on the package, boil the buckwheat until al dente. Drain, place back in the pan, and combine with the kale, onion, garlic, and dulse and season with some salt and pepper. Partly cover the pan with a lid and let rest.

5. To make the caramelized apple, halve, core, and thinly slice the apple. Heat a small splash of olive oil in a skillet and sauté the slices on each side until golden.

6. To make the seaweed topping, toast the pistachios in a skillet. Combine the pistachios with the Danish smoked salt, dulse, sea lettuce, and nutritional yeast.

7. To assemble the salad, combine the kale and buckwheat mixture with the apple slices and seaweed topping, and garnish with the fresh parsley and watercress.

ZUCCHINI SPAGHETTI with SEA PESTO

SERVES 4

This green salad with sea pesto is perfect for those balmy summer evenings: It's light and fresh. It also makes for a splendid lunch on the go or side dish for a summer picnic or barbecue. You will need a mandoline, julienne peeler, or spiralizer to cut the zucchini into nice long spaghettilike strings. If you're making this dish during the summer, go for summer purslane. It's rich in vitamin E and is the biggest source of omega-3 fatty acids of all leafy vegetables. Another great option is broccoli rabe, which will guarantee you get a healthy dose of vitamin C.

2 small baby zucchinis (as young as possible)

1½ cups (70 g) purslane or broccoli rabe, rinsed

2 tablespoons olive oil

Juice of ½ lemon

¼ cup Pesto from the Sea (see page 54)

Freshly ground black pepper, to taste

1. Rinse the zucchini and grate or slice them into spaghettilike strings.

2. Mix the zucchini strings with the purslane, olive oil, and lemon juice. Spoon in the pesto, making sure the zucchini is thoroughly coated. Season with the pepper.

CAULIFLOWER RICE with SEA ASTER

SERVES 4

Cauliflower rice! If you've never heard of it, a taste of this salad will be a delight! What makes cauliflower rice so great? First of all, you're eating an entire plate of vegetables without even thinking about it. Secondly, cauliflower rice is a great gluten-free substitute in all kinds of recipes that normally require rice or pasta. Thirdly, pairing it with sea aster creates a match made in heaven.

1 small cauliflower, rinsed and chopped

1 to 2 tablespoons extra virgin olive oil

Juice of ½ lemon

Coarse sea salt and freshly ground black pepper, to taste

½ cup (60 g) pumpkin seeds

Olive oil for sautéing

1 onion, finely chopped

1 garlic clove, minced

3.5 ounces (100 g) sea aster, rinsed and dried

Watercress, for garnish

1. In a food processor, pulse the cauliflower to a ricelike consistency. Season with olive oil, lemon juice, and salt and pepper.

2. Toast the pumpkin seeds in a hot skillet, stirring constantly until golden brown, about 5 minutes.

3. Heat a splash of olive oil in a skillet. Sauté the onion for 3 minutes. Add the garlic and sauté for 1 minute. Add the sea aster and sauté for 3 minutes, stirring continuously. Remove from the heat.

4. Mix the cauliflower rice with the pumpkin seeds and sea aster. Garnish with watercress.

MARK KULSDOM

MARK KULSDOM is a cultural historian, documentary filmmaker, and cofounder (with Lisette Kreischer, coauthor of this book) of The Dutch Weed Burger (TDWB).

What is The Dutch Weed Burger?
It's a 100 percent plant-based burger that uses kombu as a vital part of its flavor. The patty is paired with a crunchy bun enriched with chlorella, plus vegan mayo made with sea lettuce. TDWB exemplifies the many possibilities of plant-based cuisine! Seaweed is tremendously healthy, rich in essential nutrients, and provides high-quality proteins. But the taste of seaweed can be a challenge for some people—many of us aren't accustomed to eating it. TDWB subtly incorporates seaweed into a soy-based patty. Especially in its "home" territory, the Netherlands, the combination of *Dutch* and *weed* turned out to be golden. It has a rebellious ring to it, a free-spiritedness—people like to identify with it, vegan or not.

Can you describe some of the challenges of using seaweed?
There are technological challenges for large-scale cultivation of seaweed, which I think will

be met. There's also a lack of understanding about how seaweed protects coral reefs. Also keep in mind that cultivating seaweed requires neither freshwater nor farmland. Twenty years from now we'll need to know how to purify water and bring vital minerals that were washed away back into the food chain. All this while integrating seaweed into our daily diet. I'm positive and I believe that we can really do this.

So what will it take to bring seaweed to the market?
To begin with, it's important for people to get used to the flavor. Then we have to spark curiosity. That is precisely what we're doing with TDWB. We offer something new, something that intrigues people. We often see this at the pop-up stands we operate at music and other festivals. A die-hard meat eater walks by, reads our sign, THE DUTCH WEED BURGER, and becomes curious. We invite people to go on a culinary adventure, and make sure not to

give them the impression that our burger is an alternative, a second choice. And we're succeeding! TDWB is now being sold in eateries across the Netherlands and increasingly at eateries across Europe. Seaweed has a certain mystique. People are willing to give it a try. The fact that it grows in water fascinates: It's new, healthy, and from the ocean.

How do you like your seaweed?
Freshly harvested and used in a Dutch Weed Burger, of course. I eat three of them a week!

For a homemade version of The Dutch Weed Burger, see page 138.

ENTRÉES

SEITAN STEAKS with SEA ASTER & CREAMY CASHEW SAUCE

SERVES 4

Chic, surprising, and satisfying, this dish is great for a dinner party with people you love. Sea aster is increasingly finding its way into health stores and farmers' markets and onto our plates. This beautiful briny, leafy vegetable, also known as "marsh rosemary," grows near the sea and is often spotted in coastal Rhode Island. When sea aster blossoms, you'll see beautiful purple-lilac flowers that look a lot like lavender, which inspired the touch of real lavender flowers in this recipe. Bon appétit!

CREAMY CASHEW SAUCE

1 cup (150 g) unsalted cashews (soaked in water for 1 hour, then drained)

⅓ cup (100 ml) fresh water

1 tablespoon nutritional yeast

Juice of ½ lemon

1 teaspoon smoked paprika (pimentón)

Sea salt and freshly ground pepper, to taste

OVEN-ROASTED ROSEMARY POTATOES

1 pound 10 ounces (750 g) potatoes, unpeeled, rinsed, and diced

1 large onion, sliced

½ lemon, thinly sliced

2 tablespoons fresh rosemary

Coarse sea salt and freshly ground black pepper, to taste

Olive oil for roasting

SEITAN WITH SEAWEED CRUNCH

4 seitan steaks (600 g)

1 x Tofu with Seaweed Crunch (page 73; omit the tofu and replace the aonori with persil de la mer or other seaweed seasoning; see page 50)

SEA ASTER

Olive oil for panfrying

1 large onion, diced

10.5 ounces (300 g) sea aster, rinsed

Splash of white wine

Freshly ground black pepper, to taste

EXTRAS

Olive oil for frying sage

12 sage leaves

1 teaspoon dried lavender flowers

1. To make the cashew sauce, place the drained cashews, water, nutritional yeast, lemon juice, and paprika in a food processor or high-powered blender. For this recipe, it's important to grind the nuts very finely. Blend until you have a smooth, creamy sauce. You may need to scrape down the sides of the blender and mix thoroughly. Season with salt and pepper.

CONTINUES →

2. To make the potatoes, preheat the oven to 400°F (200°C). Line a baking sheet with parchment paper. Combine the potatoes with the onion, lemon slices, and rosemary and season with salt and pepper. Toss a good splash of olive oil into the potato mix and arrange over the baking sheet. Lower the oven temperature to 350°F (180°C). Bake the potatoes for 20 to 25 minutes, until golden brown and crispy. Flip them halfway through so they brown evenly.

3. To make the seitan steaks, follow the recipe for Tofu with Seaweed Crunch, but replace the tofu with seitan and the aonori with persil de la mer. You want to leave the seitan as whole pieces and not in cubes. Wait to fry the steaks until the potatoes are nearly done. When ready, fry the seitan in hot oil until crispy on each side.

4. To make the sea aster, heat a splash of olive oil in a skillet. Sauté the onion for 3 minutes. Add the sea aster and sauté while stirring for about 3 minutes. Add a splash of white wine and the aster and keep stirring until the liquid is evaporated. Remove from the heat. Season with pepper.

5. Heat a thin layer of olive oil in another skillet. Fry the sage leaves until golden brown and crispy. Drain them on some paper towels.

6. Remove the potatoes from the oven. Serve on a plate with a warm steak, some sea aster, and a scoop of creamy cashew sauce topped with dried lavender flowers. Garnish with the fried sage leaves.

TIP: For finely grinding nuts for the creamy cashew sauce (or the chlorella sauce on page 148), we recommend a Vitamix or Blendtec blender. These are also excellent for making smoothies or creamy purees.

SEAWEED GNOCCHI with SPINACH CREAM SAUCE & CARAMELIZED CHERRY TOMATOES

SERVES 4

Gnocchi are a traditional Italian pasta made from potatoes or semolina (*gnocchi alla romana*). You can find precooked gnocchi in almost every supermarket, but it's fun to make them yourself. They require some practice, so make sure to allow yourself enough time, but once you get the hang of it you will surely appreciate the homemade kind! You can also make the gnocchi ahead of time and store them in the freezer. This is a basic gnocchi recipe that we've jazzed up with chlorella and sea parsley. In this recipe, we pair the gnocchi with a creamy spinach sauce and sweet-yet-savory tomatoes for extra umami!

GNOCCHI
1 pound 10 ounces (750 g) potatoes, unpeeled and rinsed

1 teaspoon + pinch of sea salt

2½ cups (300 g) pasta flour (type 00)

1½ teaspoons chlorella

1 tablespoon + 1½ teaspoons persil de la mer (seaweed seasoning; see page 50)

¼ teaspoon freshly ground black pepper

SPINACH CREAM SAUCE
Olive oil for sautéing

2 shallots, diced

1 garlic clove, minced

1¾ pound (800 g) fresh spinach, rinsed and dried

1 cup (250 ml) almond, soy, rice, or oat creamer

2 tablespoons nutritional yeast

½ cup (15 g) fresh basil

⅓ cup (10 g) fresh mint

Fine sea salt and freshly ground black pepper, to taste

CARAMELIZED TOMATOES
Olive oil for sautéing

1 pound (500 g) cherry tomatoes, halved

2 tablespoons balsamic vinegar

EXTRAS
Nutritional yeast

Persil de la mer (seaweed seasoning)

1. **To make the gnocchi,** place the potatoes in a pan filled with ample water. Add a pinch of salt and bring to a boil. Boil for 20 to 30 minutes until the potatoes are cooked through. Drain the water and peel the potatoes with your hands (be careful—they will be very hot, but don't let them cool off). Place them in a big bowl and mash. Sift the flour over the puree and add the chlorella, persil de la mer, and salt and pepper to taste. Mix well with your hands and knead into a smooth and slightly sticky ball of dough. Cover with a towel and let rest for 10 minutes.

CONTINUES →

2. Cut the dough into quarters and cover again to prevent dehydration. Grab a dough quarter and roll into a long thin rope about 1 inch (2.5 cm) in diameter. Cut rings of 1.5 cm each. If you have a traditional gnocchi board, roll the gnocchi over the board. If not, use a fork to create those typical gnocchi ridges; with your thumb, push each gnocchi over the fork. The gnocchi will be slightly round. Place the gnocchi on a baking sheet lined with parchment paper and cover with a towel. Repeat the process for all of the dough.

3. Bring a large pot of water to a boil. Cook the gnocchi in batches, as they need to have ample space around them—they expand a little. The gnocchi are ready once they start floating. Take them out of the pan immediately with a slotted spoon and place on a baking tray lined with parchment paper or on a wooden cutting board.

4. To make the spinach sauce, heat a splash of olive oil in a skillet. Sauté the shallots for about 3 minutes, until golden brown. Add the garlic and sauté for about 1 minute. Add the spinach, stir well, and allow to wilt and for the liquid to evaporate. Add the creamer and the nutritional yeast and sauté for 2 minutes, stirring continuously. Remove from the heat. Combine the spinach, basil, and mint in a blender or food processor and process until smooth. Season with salt and pepper. Return to the pan.

5. To make the tomatoes, heat a splash of olive oil in another skillet. Add the tomatoes and balsamic vinegar and sauté over medium heat for about 3 minutes, until the tomatoes begin to caramelize. Remove from the heat.

6. Add the gnocchi to the sauce and warm over low heat. Serve with caramelized tomatoes and garnish with some persil de la mer and extra nutritional yeast.

PASTA CARBONARA with 'WEED SAUCE & TEMPEH BACON

SERVES 4

A creamy and cheesy sauce; briny, crispy vegetables; artichoke hearts sautéed in garlic and lemon; and tempeh bacon with sun-dried tomatoes all tossed with pasta *al dente* become a refreshing variation on the well-known pasta carbonara.

2 small artichokes

14 ounces (400 g) penne

Olive oil for sautéing and garnish

2 garlic cloves, minced

14 ounces (400 g) samphire, rinsed and dried

10 sun-dried tomatoes in oil, roughly chopped

Zest and juice of 1 small lemon

1 x Cheesy 'Weed Sauce (page 60)

1 x Tempeh Seaweed Snack (page 79; without the aonori and prepared while the pasta is cooking)

Fine sea salt and freshly ground black pepper, to taste

Handful of fresh parsley, finely chopped

1. Rinse the artichokes, remove the outer leaves and the fuzzy center until you get to the hearts, and chop the hearts.

2. Heat a splash of olive oil in a skillet over medium heat and add the garlic. After a minute, add the artichoke hearts and samphire. Add the sun-dried tomatoes and the lemon zest and juice. While stirring, add the Cheesy 'Weed Sauce. Allow to simmer over low heat. Turn off the heat when the artichokes are ready, about 8 minutes, though it depends on their size. Season with salt and pepper.

3. Cook the pasta in ample water with a pinch of salt until *al dente*.

4. Prepare the Tempeh Seaweed Snack, omitting the aonori to make the tempeh bacon variation.

5. Drain the pasta. Fold in the creamy artichoke and samphire sauce, serve with tempeh bacon, and garnish with a dash of olive oil and parsley.

AUTUMNAL WILD RICE SALAD with HIJIKI

SERVES 4

This packed and colorful salad is perfect for autumn. The sweet character of walnuts, sweet potatoes, oyster mushrooms, and dried cranberries nicely complement the briny hijki and dark wild rice. If you can find fresh hijiki, that's even better. If you do find it, you need only briefly blanch it in salted water to reduce the arsenic content (see page 24 for more about safe consumption). The hijiki will become bright green! As an alternative, we recommend using arame.

SALAD

2 cups (300 g) wild rice

0.5 ounce (15 g) dried hijiki or arame

2 large sweet potatoes, unpeeled, rinsed, and cubed

1 large red onion, sliced

½ orange, thinly sliced

1 heaping teaspoon smoked paprika (pimentón)

4 sprigs fresh thyme

Coarse sea salt and freshly ground black pepper, to taste

Olive oil for roasting and sautéing

7 ounces (200 g) oyster mushrooms, wiped clean and torn in small strips

½ cup (60 g) dried cranberries

1 cup (100 g) walnuts, coarsely chopped

5 ounces (150 g) baby spinach

Handful of fresh parsley, finely chopped

MISO DRESSING

⅓ cup (80 ml) orange juice

¼ cup (70 g) white miso paste

2 tablespoons extra virgin olive oil

1 tablespoon + 1 teaspoon soy sauce or tamari

1 tablespoon sesame oil

2 teaspoons mirin

1. Cook the wild rice according to directions on the package (allow at least 45 minutes), drain, place the lid on the pan, and let stand for 10 minutes.

2. Soak the hijiki for 30 minutes in lukewarm water. Once it has rehydrated, drain it and dab it dry.

3. Preheat the oven to 400°F (200°C). Line a baking sheet with parchment paper. Combine the sweet potatoes, onion, orange slices, smoked paprika, and thyme and season with salt and pepper. Toss with a generous splash of olive oil. Arrange the mixture on the baking sheet and bake for 25 to 30 minutes, until golden brown and crispy. Flip everything halfway through to allow even browning on both sides. After 20 minutes, use a spatula toss in the cranberries. Return the pan to the oven for the remaining 5 to 10 minutes, until the potatoes are golden brown and crispy. Remove from the oven and cool for a bit.

4. Heat a splash of olive oil in a skillet. Sauté the oyster mushrooms until golden brown. Season with salt and pepper.

5. To make the dressing, whisk all the ingredients into a nice smooth sauce. Combine the dressing with the hijiki.

6. Toast the walnuts in a skillet until golden brown. Divide the spinach over four plates and add a scoop of wild rice to each. Serve with the roasted vegetables. Top the salad with the hijiki and the miso dressing. Garnish with the walnuts and fresh parsley.

SPRINGTIME UDON & ARAME SEA FEAST

SERVES 4

This pasta salad is a feast for the taste buds: fresh, sweet, salty, and umami with a bite of bitterness all combined in one dish!

FRESH LIME DRESSING

Zest and juice of 1 lime

2 tablespoons safflower oil (or extra virgin olive oil)

2 tablespoons soy sauce or tamari

2 tablespoons white tahini

1 tablespoon water

1½ teaspoons mirin

1 garlic clove, minced

1 cm fresh ginger, finely chopped

½ cup (10 g) cilantro, finely chopped

Small handful of fresh mint, finely chopped

CARAMELIZED CARROTS

One 14-ounce (400 g) bunch young carrots

¾ cup + 1 tablespoon (200 ml) soy sauce or tamari

2 tablespoons + 1½ teaspoons sesame oil

Splash of mirin

NOODLES

2 tablespoons coconut oil

1 tablespoon cumin seeds

1 heaping teaspoon paprika

1 heaping teaspoon ground turmeric

1 large white onion, diced

3 to 4 tablespoons water

7 ounces (200 g) green beans, halved widthwise

1 cucumber, cubed

0.35 ounce (10 g) dried arame, rinsed and dried

9 ounces (250 g) udon noodles

za'atar, to taste

EXTRAS

⅓ cup (60 g) unsalted cashews, coarsely chopped

1 tablespoon nigella seeds

½ cup (10 g) cilantro or fresh mint, optional

1. To make the dressing, combine all the dressing ingredients and whisk until the sauce is smooth.

2. To make the carrots, halve lengthwise, or quarter the carrots if they are very big. Place in a skillet with the soy sauce, sesame oil, and mirin and bring to a gentle boil. Lower the heat, cover with a lid, and let simmer until the carrots are tender and begin to caramelize, 20 to 25 minutes. Check occasionally to make sure they don't dry out or stick to the bottom of the pan. Add water if necessary.

3. To make the noodles, heat a scoop of coconut oil in a skillet. Add the cumin, paprika, and turmeric. Stir well. Add the onion and sauté until golden brown, about 3 minutes. Add the green beans to the onions and cook for 1 minute. Add 3 tablespoons of water to deglaze the pan, and add another tablespoon if needed. Partially cover the pan with a lid, lower the heat, and let stew for 4 minutes. Add the cucumber and cook for another 4 minutes, until the juices have evaporated, the beans are crispy, and the cucumber cubes are crispy and warm. Stir in the za'atar and arame and remove from the heat.

4. Prepare the noodles according to the directions on the package. Drain and mix the vegetables into the noodles.

5. Toast the cashews and nigella seeds in a dry skillet.

6. Serve the noodles with the lime dressing and caramelized carrots. Garnish with the toasted nuts and fresh cilantro or mint, if desired.

POLENTA FRIES with SEA LETTUCE CHIPS & FRESH ASPARAGUS

SERVES 4

When green asparagus appears in stores, it's time to celebrate spring! After all those wintry root vegetables, it's a relief to savor fresh, crisp asparagus, and what better to pair it with than summery seaweeds? This recipe requires some advanced planning, so read carefully before you start. It may be difficult to find fresh dulse and sea lettuce when you want to make this at the beginning of spring. They are usually easier to find by the time the water has warmed up a bit. Frozen or pickled varieties will suffice. Allow frozen seaweed to thaw completely before preparing, and in the case of pickled seaweed, make sure to thoroughly rinse out the brine. You'll want to have on hand some Sea Aioli (page 62) or Ship Ahoy Piccalilli (page 82), which go very well with these fries!

POLENTA FRIES

Oil and flour to grease a baking dish

2¾ cup (650 ml) vegetable stock

½ garlic clove, minced

1 cup (150 g) polenta

1 tablespoon dulse flakes

1 tablespoon nutritional yeast

1 tablespoon sea lettuce flakes

1 teaspoon curry powder

1 teaspoon paprika

Sea salt and freshly ground black pepper, to taste

Olive oil for baking

1 tablespoon dried rosemary, finely chopped

Extra coarse sea salt

SEAWEED CHIPS

1.75 ounces (50 g) fresh dulse, rinsed and dabbed dry

1.75 ounces (50 g) fresh sea lettuce, rinsed and dabbed dry

Olive oil for toasting

1 tablespoon + 1½ teaspoons nutritional yeast

1 teaspoon smoked paprika (pimentón)

Freshly ground black pepper, to taste

ASPARAGUS

20 thin asparagus spears

Olive oil for sautéing

Juice of 1 lemon

Freshly ground black pepper, to taste

1. To make the polenta fries, grease a 9 x 13-inch (22 x 33 cm) baking dish with oil and dust with some flour to prevent the polenta from sticking.

2. Bring the vegetable stock and garlic to a boil in a stockpot. Add the polenta little by little while continuously stirring to avoid clumping. Lower the heat and add the dulse, nutritional yeast, sea lettuce, curry powder, and paprika. Keep stirring until the polenta thickens, 10 to 15 minutes. Season with salt and pepper. Pour the polenta into the baking pan and let cool for about an hour.

CONTINUES →

3. Preheat the oven to 400°F (200C°). Line a baking sheet with parchment paper. Once it has cooled, cut the polenta into matchsticks. Sprinkle with some oil and the rosemary. Arrange them evenly on the baking sheet, leaving some space between them so they don't stick together. Bake the polenta about 20 minutes, until golden brown. Flip them and bake for another 20 minutes. The fries are ready once they are crispy. Sprinkle with salt to taste.

4. To make the seaweed chips, lower the oven temperature to 300F° (150°C). Line a baking sheet with parchment paper and arrange the seaweed on it. Sprinkle with some olive oil (not too much, or the seaweed gets too mushy), the nutritional yeast, smoked paprika, and pepper. Massage the seaweed with your hands and make sure the seaweed is completely coated. Bake for about 15 minutes, keeping a close eye on it, as the leaves can burn quickly. Remove the baking sheet from the oven as soon as the seaweed chips are crispy. Let the chips cool, and place the polenta fries back in the oven to briefly reheat.

5. To prepare the asparagus, wash it and cut off the ends. Heat some oil in a skillet and sauté the asparagus for about 3 minutes if they're thin, or 5 minutes if thick, until golden brown. Remove from the heat and sprinkle with lemon juice. Season with pepper.

6. Serve the warm polenta fries with the seaweed chips, sautéed asparagus, and a generous spoonful of Seaweed Aioli or Ship Ahoy Piccalilli, if desired.

TIP: In addition to seaweed chips, you can also easily make seaweed popcorn (though this is more of a dish for at home on the couch). Heat a layer of sunflower oil in a deep pot with a lid. Add a handful of corn kernels and immediately cover with the lid. Cook the kernels until they pop. Once almost all the kernels have popped, sprinkle with Everything Goes 'Weed Mix (page 56) and a pinch of coarse sea salt. Place the lid back on the pot so the flavors can sink in.

WAKAME NOODLES with SPICY TOFU

SERVES 4

Seaweed, noodles, and tofu have gone hand in hand for many centuries, especially in Chinese and Japanese cuisines. This recipe for spicy seaweed pasta pays homage to this classic combination. The tofu is best when you marinate it overnight, so take this into account when you're thinking about tomorrow night's dinner.

SPICY TOFU

1 pound, 2 ounces (500 g) firm tofu

¼ cup (60 ml) ketchup

¼ cup (60 ml) soy sauce

2 garlic cloves, minced

1 cm fresh ginger, finely chopped

1 teaspoon paprika

Pinch of sambal (or other hot sauce)

NOODLES

0.2 ounces (5 g) dried wakame or 1.75 ounces (50 g) fresh wakame

9 ounces (250 g) udon noodles

Coconut oil for sautéing

1 large red onion, diced

1 large garlic clove, minced

1 inch (2.5 cm) ginger, finely chopped

1 red chile pepper, seeded and finely chopped

1 large head bok choy or 2 smaller heads, chopped

1 leek, rinsed and sliced

1 tablespoon soy sauce

1 tablespoon mirin

1 teaspoon maple syrup

EXTRAS

Scallions, for garnish

Sesame seeds, for garnish

Black pepper, to taste

1. To prepare the tofu, drain and cut it into triangular wedges about 1 cm thick. Dab dry and put the tofu in a bowl for marinating. In another bowl, combine the ketchup, soy sauce, garlic, ginger, paprika, and sambal, and pour this mixture into the first bowl to coat the tofu wedges. Cover the bowl and place it in the fridge overnight or for at least a couple of hours, so the tofu can marinate. Stir occasionally. After a few hours, or the next day, strain the excess liquid, reserving it for later.

2. To make the noodles, soak the dried wakame according to the directions, drain, and dab dry. If using fresh seaweed, rinse it thoroughly and remove the brine. Chop the seaweed into medium-size pieces, about 5 mm.

3. Prepare the noodles according to the directions on the package. Heat a bit of coconut oil in a skillet. Sauté the onion until golden brown, about 5 minutes. Add the garlic, ginger, and chile and sauté for about 1 minute. Add the bok choy and leek and stir-fry until they start to soften, about 5 minutes. Add the soy sauce a little at a time, tasting to make sure it isn't too salty. Add the mirin and maple syrup. Stir-fry until the bok choy and leek are done. Remove from the heat and mix the noodles with the wakame.

4. To make the spicy tofu, heat a generous layer of oil in a skillet. Sauté the tofu wedges until golden brown on all sides.

5. Serve the tofu with the warm seaweed and vegetables and top with some of the reserved marinade. Garnish with fresh scallions, sesame seeds, and black pepper.

SEAWEED SOCCA PIZZA

SERVES 4

Socca is a traditional French savory pancake made from chickpea flour. This pancake can also be used to make pizza, and gluten-free pizza at that! You can use the basic recipe for all kinds of dishes, but for *Ocean Greens* we can't help but have umami play first fiddle and have seaweed as a primary seasoning. Best served with a side of mixed green salad.

SOCCA

3 cups (340 g) chickpea flour

½ cup (75 g) kalamata olives, pitted and sliced

1 tablespoon + 1½ teaspoons fresh Italian herbs (e.g., oregano, thyme, rosemary), finely chopped

1 tablespoon + 1½ teaspoons persil de la mer (seaweed seasoning, see page 50)

1½ teaspoons fine sea salt

½ teaspoon freshly ground black pepper

2½ cups (600 ml) lukewarm water

Juice of ½ lemon

¼ cup + 1 tablespoon olive oil, plus more for cooking

SOCCA TOPPINGS

1 ounce (25 g) sea spaghetti

16 green asparagus spears

14 ounces (400 g) cherry tomatoes, halved

1 red onion, sliced

Olive oil for roasting

Coarse sea salt and freshly ground black pepper, to taste

1 x Pesto from the Sea (page 54)

EXTRAS

Fresh basil or arugula, optional

Nutritional yeast, optional

Everything Goes 'Weed Mix (page 56), optional

Extra virgin olive oil

1. Combine the chickpea flour, olives, Italian herbs, persil de la mer, salt, and pepper in a large bowl. While whisking, add the water, lemon juice, and olive oil. Whisk until thoroughly combined. The batter should be pretty thick. Cover with a dish towel and let rest for at least 30 minutes to let it thicken.

2. Cook the sea spaghetti following the directions on the package, drain, and set aside.

3. Preheat the oven to 400°F (200°C). Line a baking sheet with parchment paper. Clean the asparagus and remove the ends. Place the asparagus with the cherry tomatoes and red onion on the baking sheet. Add some olive oil, toss, and season with salt and pepper. Roast the vegetables in the oven about 25 minutes, until golden brown and crispy. Remove the vegetables and lower the temperature to 350°F (180°C).

4. Set out a large plate for the cooked socca. Heat a splash of olive oil in a skillet. Add a scoop of the socca batter to make pancakes that are about 8 inches (20 cm) in diameter. Cook until the socca is firm, about 3 minutes; flip it halfway through to cook both sides, just as you would for regular pancakes. Slide the socca onto the plate. Repeat with the rest of the batter. You should be able to make about six pancakes.

5. Arrange each socca into a pizza: Cover a baking sheet with parchment paper. Place the socca "crust" on the paper. If you have several baking sheets or if you've made small soccas, you might be able to fit them all in the oven at once. Spread Pesto from the Sea over the soccas. Top with the oven-roasted vegetables and some sea spaghetti. Bake in the oven for 2 to 3 minutes. Garnish with fresh basil or arugula, some nutritional yeast, or Everything Goes 'Weed Mix, and finish with a drizzle of olive oil.

THE HOMEMADE DUTCH WEED BURGER

SERVES 4

The Dutch Weed Burger is a 100 percent plant-based hamburger we developed with the help of others, enriched with kombu from a system of tidal estuaries in the Dutch province of Zeeland. The burger contributes significantly to the wider introduction of seaweed in the Netherlands and abroad and has been conquering the mouths and hearts of people all over the world. A documentary was made about the burger and filmed on the streets of New York City by Mark Kulsdom (see page 114). The Dutch Weed Burger has been taking over restaurant menus across the Netherlands. To satisfy your desire at home, we present here a simpler but equally tasty variation. It is best served with mixed greens and french fries.

TEMPEH PATTIES

8 ounces (250 g) tempeh

⅓ cup (75 ml) soy sauce or tamari

2 tablespoons ketchup

2 tablespoons maple syrup

1 tablespoon vegan Worcestershire sauce

1 teaspoon smoked paprika (pimentón)

Olive oil

CONDIMENTS & TOPPINGS

0.2 ounces (5 g) dried wakame or 1.75 ounces (50 g) fresh wakame, roughly chopped

Olive oil for sautéing

1 large red onion, sliced

1 red bell pepper, thinly sliced

1 to 2 tablespoons balsamic vinegar

Coarse sea salt and freshly ground black pepper, to taste

1 x Avocado & 'Weed Hummus (page 63)

1.5 ounces (75 g) baby spinach, rinsed and dabbed dry

4 whole wheat pita breads or flatbreads (or more for double burgers)

1. To prepare the burgers, cut the tempeh into thin slices, either horizontally or vertically. In a bowl, combine the soy sauce, 1 tablespoon of ketchup, the maple syrup, Worcestershire sauce, and smoked paprika. Place the tempeh slices in a wide-mouthed shallow bowl and pour the marinade on top. Cover with a dish towel and marinate for at least 30 minutes. Flip the slices halfway through and make sure they are fully covered in marinade.

2. To make the toppings, soak the dried wakame according to directions on the package, drain, and dab dry. If you're using fresh wakame, make sure to rinse well and remove all the brine. Dab dry and roughly chop into pieces about 1½ to 2 mm.

3. Heat a thin layer of oil in a skillet and sauté the onion for about 5 minutes, until golden brown. Add the pepper strips and sauté for 5 minutes, until the pepper becomes tender. Add the wakame and sauté

EAT WEED, LIVE LONG
—MARK KULSDOM & LISETTE KREISCHER

for 2 minutes. Add the balsamic vinegar and cook until the vinegar is absorbed and excess juice has evaporated. Season with salt and pepper. Remove from the heat and let sit in the pan until serving.

4. Strain the tempeh, reserving the marinade for later. Heat a splash of olive oil in a clean skillet. Panfry the tempeh slices for 5 minutes, flipping a few times. Add the remaining tablespoon of ketchup and a splash of the marinade, and coat the tempeh. Cook until the tempeh slices are crispy on both sides.

5. Cut the pita breads in half and toast them. Spread a generous dollop of Avocado & 'Weed Hummus on one half of the pita, put some spinach and a few tempeh slices on top, and top with the wakame-pepper mixture and another scoop of hummus.

6. If you want, you can use another pita in the middle to make double Weed Burgers.

SQUASH & SEAWEED PANCAKES

MAKES ABOUT 10 PANCAKES

Nice thick pancakes filled with pumpkin and layered with seaweed pesto and fresh purslane. Doesn't that sound exciting? The sweet pumpkin combines well with the briny kombu, and together they truly explode with taste. It's nice with the Tempeh Seaweed Snack on page 79 (without the nori, as this recipe already contains enough seaweed) and a green, briny salad. Bon appétit!

PANCAKES

0.7 ounce (20 g) fresh kombu

1 small hokkaido squash or other winter squash

3 cups (350 g) spelt flour

1 tablespoon baking powder

1½ teaspoons dried rosemary

1 teaspoon paprika

½ teaspoon curry powder

1 tablespoon olive oil, plus more for cooking

1 cup (250 ml) warm water

Fine sea salt and freshly ground black pepper, to taste

EXTRAS

1 x Pesto from the Sea (page 54)

3½ cups (150 g) winter purslane, rinsed and dried

Extra virgin olive oil

Lemon juice

Everything Goes 'Weed Mix (page 56)

1. Thoroughly rinse the kombu to remove all the brine. Dab dry and chop. Peel the squash, remove the seeds, and cube the flesh.

2. Bring a pot of water to a boil and add the squash cubes; boil until tender, about 8 minutes. Drain and blend in a food processor into a smooth puree. Separate 1⅔ cups (400 g) of puree for the pancakes. If there's extra, you can freeze it for later.

3. Combine the flour, baking powder, rosemary, paprika, and curry powder in a bowl. In another bowl, mix the squash puree with the oil, kombu, and warm water and then add this to the flour. Whisk thoroughly until you have a thick, smooth pancake batter. Season with salt and pepper.

4. Heat a splash of olive oil in a skillet and pour some batter in the pan. Rotate the pan so you get a small, thick, and round pancake about 4 inches (10 cm) in diameter. Cook until golden brown on both sides, about 6 minutes, flipping after 3 minutes. You can use a greased ring in the pan so all pancakes are evenly sized and perfectly round. Keep the pancakes warm by setting them on a plate that's resting on a pot of simmering water.

5. Top the pancakes with Pesto from the Sea and fresh winter purslane. Sprinkle the purslane with some olive oil and lemon juice and garnish with Everything Goes 'Weed Mix.

PRANNIE RHATIGAN

PRANNIE RHATIGAN is a medical doctor, born on the northwest coast of Ireland where she still lives and works. The author of *Irish Seaweed Kitchen,* she gives lectures and workshops and regularly takes people on seaweed walks.

When did you start cooking with seaweed?

I've always been around seaweed. I remember watching my parents harvest seaweed near our house in Ireland as a child. My father, who was also a doctor, knew the value of what could be found along the shoreline. But there has long been a stigma attached to it as well—seaweed was a last resort: You only ate it if you were poor. Our neighbors used to tease him, saying, "Times must be hard if you're down on the shore." The Irish tradition of eating seaweed is very old—there's a seventh-century law which stipulated that any traveler calling at one's door should be offered a serving of dulse.

After graduating from medical school and returning to my hometown, I was occasionally asked for seaweed advice because of my family. Soon, the local Organic Centre approached me to do courses. The students were curious about seaweed recipes and as it happened my mother had loads of them. I suppose one thing led to another and after six years of collecting, writing, and experimenting there was a cookbook.

Can you tell a bit more about this experimenting?

People would say, "For you, it's easy; you grew up eating seaweed. But what about us?" In response I began asking them for their absolute favorite recipe and tried to find a way to incorporate seaweed into those dishes. We organized lots of tasting parties and blind taste tests. I would sneak seaweed into things like bread—a very traditional way of incorporating seaweed—and ask people to serve it to their unwitting colleagues or family members. The umami of seaweed will lift any dish, as long as it's properly balanced. The many positive reactions meant we were onto something.

So it's about convincing people through their taste buds?

Yes, and about educating them. That's what we do on our seaweed identification and harvesting walks. We go out on the days around the full and new moons. That's when the tide goes out extremely far, and a garden of seaweed is exposed for an hour or so. It's superb to see the different species, the kelps, the reds, and the browns, all arranged in their own zones along the shore. All we have to do is use our scissors and give the ones we want a little haircut. But there's no point in harvesting something if you don't know how to use it in the kitchen, so I combine tasting and identification during my walks. Recently, I had a participant from Japan who was perplexed that nori actually grows in the wild out here. Imagine: Irish rocks covered in dark chocolate-like seaweed.

SWEET & SAVORY

SAVORY SEAWEED MUFFINS

MAKES 8 TO 10 MUFFINS

Seaweed is delicious in muffins, especially savory muffins, served with a dollop of Sailor's Butter (page 58). These umami delicacies are great for a special breakfast, as a snack, or with dinner instead of bread.

1 cup (250 ml) oat or almond milk

⅓ cup (100 ml) olive oil

Zest and juice of ½ lemon

2 cups (250 g) whole grain spelt flour

1 tablespoon dried sea lettuce

1 tablespoon dulse flakes

1 tablespoon nutritional yeast

1½ teaspoons baking powder

1½ teaspoon dried thyme

1 teaspoon fine sea salt

1 teaspoon smoked paprika (pimenton)

6 sun-dried tomatoes in oil, sliced

2 scallions, sliced

Toasted sesame seeds

1. Preheat the oven to 375°F (190°C). Line a silicone muffin pan for at least eight muffins with paper muffin cups.

2. Combine the milk, oil, and lemon juice in a bowl. Allow the mixture to rest for about 5 minutes (the milk will begin to clump up—that is what we want).

3. Combine the flour, dulse, sea lettuce, nutritional yeast, baking powder, thyme, salt, smoked paprika, tomatoes, scallions, and lemon zest in a separate bowl. Mix thoroughly.

4. Stir the liquids into the solids, and mix until any lumps have disappeared but not any longer. It is vital that the baking powder isn't activated before it's in the oven. If you stir too long, the baking powder will begin to work.

5. Divide the batter among the muffin cups, top with toasted sesame seeds, and place in the oven. Bake the muffins for about 25 minutes, until golden.

SAILOR'S BREAD

Sailor's bread is technically more of a cake than a bread, as there is no yeast involved. You could, however, serve this recipe as bread: For lunch, a slice is delicious with salad, or for dinner, it can accompany of a bowl of soup. It goes well with a scoop of tomato jam or Sailor's Butter (page 58).

0.2 ounce (5 g) dried wakame

½ cup (75 g) unsalted cashews

3 cups (350 g) spelt flour

⅓ cup (20 g) nutritional yeast

1 tablespoon baking powder

1 teaspoon Danish smoked salt

Pinch of freshly ground black pepper

Zest and juice of ½ lemon

2 cups (450 g) unsweetened soy yogurt

½ cup (120 ml) olive oil

EXTRAS
Nutritional yeast, optional

Everything Goes 'Weed Mix (page 56), optional

1. Soak the dried wakame according to the directions on the package, drain, and dab dry.

2. Grind the cashews in a food processor to a coarse crumble.

3. Preheat the oven to 400°F (200°C). Grease a 9 x 5-inch (23 x 13 cm) bread pan with oil and sprinkle with some flour.

4. Combine the ground cashews, flour, nutritional yeast, baking powder, salt, pepper, and lemon zest in a bowl.

5. In another bowl, combine the wakame, soy yogurt, oil, and the lemon juice. Stir this mixture into the flour mixture until any lumps have dissolved but not any longer, so as not to activate the baking powder before it's in the oven. Don't stir too long, or the baking powder will activate.

6. Pour the mixture into the greased baking pan and bake for about 50 minutes, until golden brown. The bread is done when you can stick in a knife or toothpick and it comes out clean. Remove the bread from the oven and allow it to cool before you take it out of the pan. Garnish with some nutritional yeast or Everything Goes 'Weed Mix, if desired.

DECADENT CHLORELLA & BLACKBERRY YOGURT

SERVES 4

Chlorella is sometimes called a "superfood" by nutritionists—a coveted title it owes to its enormous amount of protein, chlorophyll, and omega-3 fatty acids. And you can taste it. Chlorella has an intense flavor, which you need to learn how to work with in the kitchen. In powder form, it's a wonderful addition to smoothies. Always begin with just a pinch per person, and increase the serving to taste and as you become more familiar with this power food. In this recipe, chlorella is used in a special almond sauce that you can serve with all kinds of dishes. The sauce is delicious with pancakes or on a sorbet. Making the almond milk yourself is an added bonus; just give it a try. Almond milk is extremely creamy and rich in vitamins and minerals—the perfect companion for this green superfood!

CHLORELLA SAUCE
1 heaping cup (150 g) almonds (raw or blanched)

2 cups (500 ml) water

Zest of ½ lemon

Pinch of salt

Pinch of vanilla powder (or vanilla seeds from 1 bean)

1 tablepoon chlorella powder

YOGURT
2½ cups (560 g) unsweetened soy or coconut yogurt

2 bananas

Juice of ½ lemon

1¾ cups (250 g) blackberries

1½ teaspoons maple syrup

EXTRAS
4 teaspoons hemp seeds

Fresh mint leaves

1. To make the chlorella sauce, steep the almonds in water for 8 to 10 hours (or overnight). Drain and put them in a high-powered blender with the water, lemon zest, salt, and vanilla powder. Blend at high speed until it's nice and smooth. Blend in the chlorella and pour the contents into a glass bottle with a lid. You can preserve this sauce up to four days in the fridge; you will have extra! Shake well before use.

2. To make the yogurt, blend the yogurt with the bananas and lemon juice until creamy. Separately, blend the blackberries with the maple syrup into a bright purple sauce. Scoop some of the blackberry sauce over a bowl of the yogurt and finish with 2 to 3 teaspoons of the chlorella sauce per person. Garnish with some hemp seeds and mint leaves.

TIP: You can also make almond milk from this recipe! Follow the steps for the chlorella sauce, but before blending in the chlorella, pour the almond puree through a cheesecloth. Store the collected milk in a bottle.

GREEN KICK START SMOOTHIE BOWL

SERVES 4

There's perhaps a little too much green in this recipe, but did you ever hear Popeye complain about that? This breakfast offers an enormous kick start at the beginning of the day. Thanks to the algae, "milk," spinach, and banana, you'll receive a healthy dose of important nutrients (protein, iron, and fiber) that will make you feel like Popeye for the rest of the day.

2 tablespoons crushed flaxseeds

¼ cup (60 ml) boiling water

4 bananas, chopped

5 ounces (150 g) spinach

2 cups (450 ml) almond, rice, spelt, or soy milk (unsweetened)

Juice of 2 oranges

1 tablespoon + 1 teaspoon spirulina powder

EXTRAS

4 teaspoons hemp seeds

¼ cup (60 g) soy or coconut yogurt

4 teaspoons chlorella sauce (page 148), optional

1. Put the flaxseeds in a cup or bowl and pour the boiling water over them until they are just submerged. Let soak for 5 minutes.

2. Put the bananas, spinach, milk, and orange juice in a high-powered blender or food processor (if it doesn't fit, halve the recipe and blend in batches). Add the spirulina and soaked flaxseeds, including the water. Blend at high speed into a wonderful green smoothie.

3. Serve with some hemp seeds and a scoop of soy yogurt, and garnish with a teaspoon of chlorella sauce per person, if desired.

SPIRULINA & STRAWBERRY POPSICLES

MAKES 6 POPSICLES

This is a very simple recipe. Mix, blend, pour, freeze, and eat. These Spirulina & Strawberry Popsicles are ideal to help you cool down in the summer and make for a very easy way to get your vitamins while snacking. Ideal for the kids, even more so if they're allowed to make them themselves. They will love it!

1½ cups (350 ml) coconut milk

1 banana, chopped

¼ cup (50 ml) rice syrup

1 heaping teaspoon spirulina powder

1 cup (150 g) strawberries

Pinch of vanilla powder (or vanilla seed from 1 bean)

1. Blend the coconut milk, banana, rice syrup, and spirulina until smooth. Pour this mixture into a bowl.

2. Blend the strawberries and the vanilla powder into a smooth red sauce.

3. Divide the coconut-banana mixture among six silicone ice-pop molds. Swirl the strawberry sauce into the mixture and insert the popsicle sticks. Freeze for at least 6 hours.

EASY CHLORELLA ICE CREAM

SERVES 4

This may be the easiest-to-make ice cream on the planet, and possibly the greenest, too. You can make this recipe with all kinds of fruit, but since chlorella mixes well with banana, that forms the basis for this ice cream. Do you want to go for a very colorful combination? Replace the banana with strawberries. Can you guess what color the ice cream will become?

4 bananas, chopped

¼ cup (50 ml) applesauce

2.5 ounces (75 g) dark chocolate

¾ cup + 1 tablespoon (200 ml) almond, soy, rice, or oat creamer

⅔ cup (20 g) fresh mint leaves, plus more to garnish

2 teaspoons chlorella powder

Pinch of vanilla powder (or vanilla seed from 1 bean)

A few raspberries or red berries (optional)

Vegan whipped cream (optional)

1. Put the banana pieces in a plastic container and scoop the apple sauce into an ice-cube tray. Freeze both for a couple of hours.

2. Meanwhile, finely chop the chocolate.

3. Once the bananas and applesauce are frozen, remove them from the freezer and allow to thaw slightly. Put the creamer, mint leaves, chlorella, vanilla, and slightly-less-than-frozen banana and applesauce in a high-powered blender or food processor and blend until thoroughly combined. It should be thick and creamy.

4. Stir in nearly all the chocolate chunks and serve immediately. Garnish with fresh mint and the rest of the chocolate, as well as red berries and vegan cream, if using.

JUST ANOTHER TIRAMISU

SERVES 4

No green seaweed this time but agar, an ingredient you may have heard about before—and may even have used. Agar, or agar-agar, is a thickener that works similarly to gelatin, but it's plant-based: made from seaweed. Once you have gotten the hang of how to apply agar in your cooking, you will no doubt begin to value this versatile and useful thickener. It's especially useful for whipping up mousses and crèmes, particularly this delicious coconut mousse, dressed as a fashionable tiramisu.

10 Medjool dates, pitted

½ cup (125 ml) strong, freshly brewed coffee (from grounds preferred)

2⅓ cups (550 ml) coconut milk

⅓ cup (80 ml) maple syrup

Zest of ½ lemon

2 tablespoons (15 g) agar powder

Splash of water

1 teaspoon vanilla powder (or the seeds of 1 bean)

Pinch of salt

8 (round) whole wheat tea biscuits, crumbled

Cocoa powder, for garnish

1. Soak the dates in the coffee for 10 minutes.

2. Combine the coconut milk, maple syrup, and lemon zest in a saucepan over low heat. In a small bowl, add a splash of water to the agar and, while stirring, pour the emulsion into the coconut mixture. Increase the heat a little and bring to a boil while stirring continuously. Lower the heat again and partly cover with a lid. Let simmer for 10 minutes, stirring occasionally to prevent it from curdling or sticking to the pan. Remove from the heat, allow to cool, and set aside in the fridge for at least 1 hour.

3. Place the coffee, dates, vanilla powder, and salt in the food processor and puree until thick and smooth.

4. Once the coconut mousse has stiffened completely, remove it from the fridge. Process it with an immersion blender until smooth and creamy.

5. To assemble, add a scoop of coconut mousse to a small glass, sprinkle with some biscuit crumbs, and top with a scoop of the date-coffee sauce. Repeat. Another scoop of mousse, some biscuit crumbs, and another scoop of sauce on top. Finish by adding one final small scoop of mousse, then dust with some cocoa powder. Allow the tiramisu to stiffen in the fridge for at least 15 more minutes before serving.

CHOCOLATE CHIP & 'WEED COOKIES

MAKES 10 TO 12 COOKIES

This book presents no shortage of challenges when it comes to flavors and cooking: seaweed ice cream, seaweed bread, green breakfast smoothies, and now it's chocolate and seaweed in a cookie! This is exactly what seaweed does; it challenges, tickles, intrigues, and seduces. Learning how to taste, mix, and comfortably handle these vegetables from the sea are the keys to this seaweed adventure. Hopefully, seaweed will secure a permanent spot in your pantry.

2 cups (225 g) spelt flour

2.5 ounces (75 g) dark chocolate, chopped

½ cup (50 g) pecans, roughly chopped

¼ cup (50 g) unrefined cane sugar

1 tablespoon dried sea lettuce flakes

1 teaspoon vanilla powder or extract (or vanilla seeds from 1 bean)

½ teaspoon baking powder

Pinch of fine sea salt

Zest of ½ lemon

⅓ cup (100 ml) rice syrup

⅓ cup (100 ml) sunflower oil

1 tablespoon water

1. Preheat the oven to 350°F (180°C). Line a baking sheet with parchment paper. Combine the flour, chocolate, pecans, cane sugar, sea lettuce, vanilla, baking powder, salt, and lemon zest in a bowl. In another bowl, combine the rice syrup, oil, and water and then stir it into the flour mixture. Knead well with your hands until the ingredients come together in a dough.

2. Form small balls from the dough, place them on the baking sheet, and flatten them slightly (but they should still be thick).

3. Bake the cookies for 12 to 15 minutes, until golden brown. The cookies will be crispy on the edges and soft in the middle. Remove the baking sheet from the oven and allow the cookies to cool.

FESTIVE CHOCOLATE, RASPBERRY & SEAWEED CAKE

SERVES 12

In a book filled with delicious recipes, we shouldn't be missing a cake! Even a book about seaweed should have cake: a seaweed cake! You might not expect it, but seaweed and dark chocolate go very well together. Hence this fierce recipe for an even fiercer chocolate cake, suitable for gourmet cooks and foodies alike. Allow for some time, however, as the various layers that make up this cake deserve your attention. Once you've decorated it, you'll steal the show with this special and festive treat.

GANACHE

7 ounces (200 g) dark chocolate

½ cup (120 ml) almond or soy creamer

2 tablespoons plant-based margarine, at room temperature

1 tablespoon maple syrup

CHOCOLATE CAKE

2 tablespoons ground flaxseeds

¼ cup + 1 tablespoon (75 ml) boiling water

¾ cup (190 g) soy yogurt

⅓ cup (100 ml) applesauce

¼ cup (50 ml) maple syrup

2¼ cups (275 g) spelt flour

1 cup (100 g) cocoa powder

½ cup (50 g) hazelnut flour (see Tip)

1 tablespoon baking powder

2 teaspoons vanilla powder (or vanilla seeds from 2 beans)

½ teaspoon fine sea salt

Zest of ½ lemon

1 sheet nori, crumbled in small pieces

7 tablespoons (200 g) plant-based margarine, at room temperature

1 cup (200 g) unrefined cane sugar

2 tablespoons hot water

VANILLA CHLORELLA CRÈME

1 cup + 1 tablespoon (225 g) plant-based margarine, at room temperature

2 cups (250 g) powdered sugar

2 tablespoons almond, oat, rice or soy milk

2 teaspoons vanilla powder (or vanilla seeds from 2 beans)

Zest of ¼ lemon

1 heaping teaspoon chlorella powder

RASPBERRY SAUCE

¾ cup (100 g) fresh raspberries

EXTRAS

1¼ cups (150 g) fresh raspberries

¼ cup (35 g) unsalted raw hazelnuts

½ sheet nori, crumbled into small pieces

Coarse sea salt, for garnish

1. To make the ganache, break the chocolate into small pieces. Heat the creamer, margarine, and maple syrup in a double boiler. Once the margarine has melted, remove the bowl from the pot of boiling water to stop heating it. Add the chocolate pieces and stir until the chocolate has completely melted. Allow to cool completely and stir before serving to loosen.

CONTINUES →

2. To make the chocolate cake, preheat the oven to 350°F (180°C). Grease a cake pan with margarine and dust with flour.

3. Put the flaxseeds in a cup or small bowl and mix with the boiling water. Let soak for 5 minutes.

4. Mix the soy yogurt, applesauce, and maple syrup in a bowl. In another bowl, combine the spelt flour, cocoa powder, hazelnut flour, baking powder, vanilla powder, salt, lemon zest, and crumbled nori.

5. Whisk the margarine in a bowl with a fork. Cream the margarine and sugar by beating with a mixer for about 4 minutes until fluffy (you can also use a food processor).

6. Add the applesauce mixture to the flour mixture and then add in the creamed margarine. Add the flaxseeds and hot water and stir until any lumps have dissolved but not any longer. It should be a firm batter. Pour into the greased pan and bake for 50 to 60 minutes, or until a knife slid into the middle comes out clean.

7. Remove the cake from the oven and allow to cool for about 10 minutes. Loosen the cake around the edges with a knife and then carefully remove it from the pan. Let cool completely. Then carefully slice the cake into two layers.

8. Meanwhile, **to make the chlorella crème,** beat the margarine in a bowl with a mixer at medium speed. Beat in the powdered sugar little by little. Increase the mixing speed and beat the margarine for about 2 minutes. Add the almond milk, vanilla, and lemon zest. Beat at high speed for 4 minutes until the sugar is completely blended into the margarine and the crème is light and creamy (you can also do this in a food processor).

9. To make the raspberry sauce, puree the raspberries into a smooth red sauce.

10. For the extras, toast the hazelnuts in a skillet until golden brown. The skin may stay on. Chop coarsely.

11. You now have three beautiful sauces to decorate the cake with: the ganache, the chlorella crème, and the raspberry sauce. It's up to you to make it beautiful. One method is to spread one half of the cake with raspberry sauce and scoop a thick layer of ganache on top. Swirl some of the chlorella crème on top. Place the other half of the cake on top. Spread the remaining ganache over the entire cake. Swirl some chlorella crème on top and garnish with roasted hazelnuts, the remaining raspberries, and crumbled nori. Sprinkle with a little bit of coarse sea salt.

TIP: You can make your own hazelnut flour by finely grinding unsalted, raw hazelnuts in a food processor.

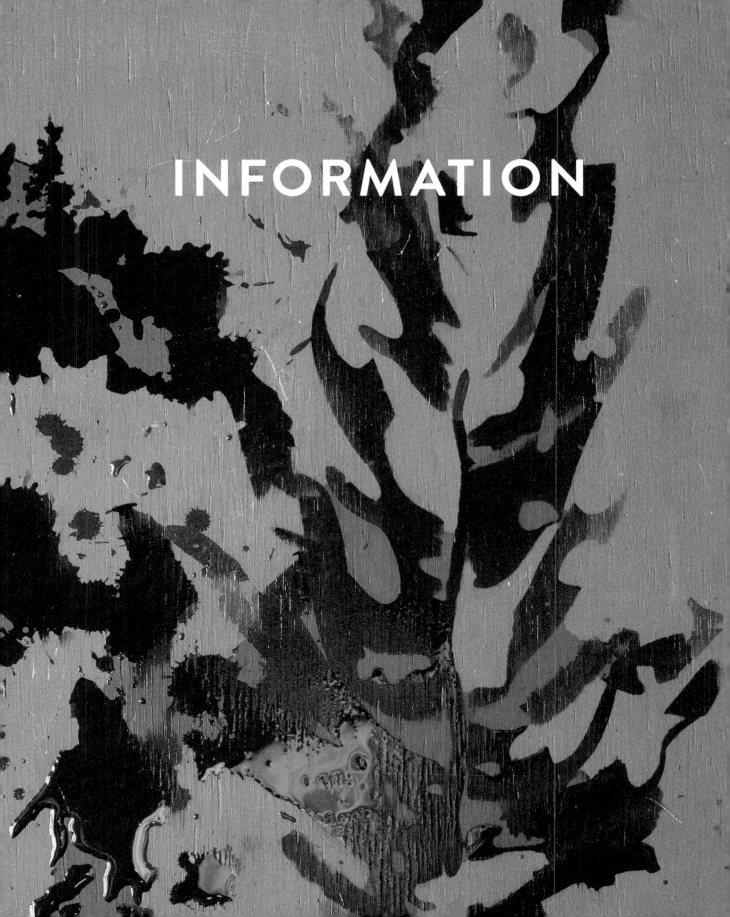

INFORMATION

GENERAL INFORMATION

ALGAEBASE
Database of information on algae, including terrestrial, marine, and freshwater organisms
algaebase.org

EAT THE WEEDS
Florida-based foraging, plant identification, and general information classes; list of foraging instructors by state
eattheweeds.com

FORAGE SF
Seaweed identification and foraging classes on the Sonoma Coast of California
foragesf.com

MAINE SEAWEED
Coastal Maine harvester, open to visitors
theseaweedman.com

NATIONAL PARK SERVICE
"Field Guide to Marine Plants/Algae" in Acadia National Park, Maine
nps.gov

THE SEAWEED SITE
Source of general information about all aspects of seaweed
seaweed.ie

THE UNIVERSITY HERBARIA AT UC BERKELEY
Collection of marine algae found on the Pacific coast
ucjeps.berkeley.edu/CPD/algal_research.html

US DEPARTMENT OF AGRICULTURE
Database of plants, including sea vegetables
plants.usda.gov

PLANTS FOR A FUTURE
A database of edible plants
pfaf.org

THE DUTCH WEED BURGER
dutchweedburger.com/en

MAINE COAST SEA VEGETABLES
seaveg.com

NORTH SEA FARM FOUNDATION
northseafarm.com

VEGGIE IN PUMPS
veggieinpumps.com

ZEEWAAR
zeewaar.nl

WHERE TO BUY SEAWEED

All products can be ordered online, unless marked otherwise, and can be shipped within the continental US.

To source fresh seaweed and sea vegetables, check your local Asian market.

ASIAN FOOD GROCER
asianfoodgrocer.com
Agar, kombu (kelp), wakame

ATLANTIC HOLDFAST SEAWEED COMPANY
atlanticholdfast.com
Dulse, Irish moss, kombu (kelp), nori, wakame

BRITTANY SEA SALT
brittanysalt.com
Persil de la mer (sea parsley), sea spaghetti

EDEN FOODS
edenfoods.com
Agar, arame, dulse, hijiki (hiziki), kombu (kelp), nori, wakame

FINE FOOD SPECIALIST
finefoodspecialist.co.uk
Aonori, dulse, nori, sea spaghetti, wakame

GREEN FOODS
greenfoods.com
Chlorella, spirulina

H MART
hmart.com
Hijiki, kombu (kelp), nori, sea lettuce, spirulina, wakame

IHB NUTRITION
ihbnutrition.com
Spirulina

IRISH SEAWEEDS
irishseaweeds.com
Devil's apron (sweet kelp), dulse, Irish moss, kombu (kelp), oarweed (*L. digitata*), sea lettuce, sea spaghetti, wakame

IRONBOUND ISLAND
ironboundisland.com
Dulse, kombu (kelp), nori, oarweed (*L. digitata*), wakame

JUST SEAWEED
justseaweed.com
Aonori, bladderwrack, devil's apron (sweet kelp), kombu (kelp)

MAINE COAST SEA VEGETABLES
seaveg.com
Dulse, fucus, Irish moss, knotted wrack (Norwegian kelp), kombu (kelp), nori, persil de la mer (sea parsley), sea lettuce, wakame

MAINE SEAWEED
theseaweedman.com
Bladderwrack, dulse, kombu (kelp), oarweed (*L. digitata*)

MENDOCINO SEA VEGETABLE COMPANY
Call and mail orders only
seaweed.net
Dulse, fucus, kombu (kelp), nori, wakame

MARX FOODS
marxfoods.com
Samphire

"SALT IS BORN OF THE PUREST PARENTS: THE SUN AND THE SEA."
—PYTHAGORAS

MARX PANTRY
marxpantry.com
Arame, hijiki, kombu (kelp), nori, samphire (pickled), sea lettuce, wakame

OCEAN APPROVED
Fresh seaweed available; order by phone
oceanapproved.com
Kombu (kelp)

OCEAN HARVEST SEA VEGETABLE COMPANY
ohsv.net
Dulse, kombu (kelp), nori, sweet kombu, wakame

SEASNAX
seasnax.com
Kombu (kelp), wakame

VITAMIN SEA MAINE SEAWEED
vitaminseaseaweed.com
Bladderwrack, dulse, Irish moss, kombu (kelp), nori, sea lettuce, wakame

WHOLE EARTH HARVEST
wholeearthharvest.com
Samphire

WHOLE FOODS
Ask in store
wholefoodsmarket.com

Index

DULSE

WAKAME

NORI

ARAME

HIJIKI

ROCKWEED

SEA LETTUCE

SEA SPAGHETTI

KOMBU (KELP)

Special Thanks

OCEAN GREENS came about thanks to a big group of people. We would like to thank:

Gerard Zwijnenberg and Willem van der Zwan

Eef Brouwers

Bert Groenendaal

The team of Kosmos Publishers (Melanie, Claire, Judith, Bert)

Ingrid Meurs

Mark Kulsdom, Marieke Zwerver and Sanne Boersma – The Dutch Weed Burger

Harry Kreischer

Astrid Kleijnjan

Pooh

Renee Jager

Jonathan van Alteren

Dos Winkel – Sea First Foundation

Willem Brandenburg – Wageningen UR

Carina Noordervliet – Seamore

Iris Rozendaal

Ester Koorn – Marc. Foods

Kevin van Reijen – DotS

Nevil van Reijen – DotS

Joost van Beek

Christel Schollaardt and Roxali Bijmoer – Naturalis

Diewertje van Wering

Katharina Andrés

Freek van den Heuvel

Marloes Kneppers

Alexander Ebbing

Frank Diependaal

Lotje van der Poll

Marlies Draisma

Job Schipper – Hortimare

Kees Boender – Your Well

Jennifer Breaton and Rebecca Wiering – Zeewaar

WAAR in Den Haag

Verzamelaars van Vrolijkheid for the beautiful vintage kitchen accessories

Alissa+Nienke for the beautiful handmade porcelain

Jennifer de Jong for her beautiful ceramics

De Spiegeltent in Gouda for shooting all those beautiful images

The crew of the MS Terschelling

Marieke Eyskoot

Geert Gordijn

Guy Buyle – Centexbel

We would also like to thank the colleagues of Schuttelaar & Partners for their advice and patience every time we disrupted their week with our enthusiasm. This office has, often invisibly, stood at the birth of dozens of initiatives, projects, and organizations. All to make the world a little healthier and greener. Seaweed farming in the North Sea and this book, *Ocean Greens*, are a testament to their work.

Authors & Organizations

LISETTE KREISCHER is the author of seven books on vegan food and cooking and living an ecofabulous way of life. She is also the cofounder of the company behind The Dutch Weed Burger, which is the subject of a feature-length documentary filmed in NYC and screened around the world. Lisette is committed to spreading the word that plant-based food is easy to make, tasty, and healthy, and belongs in everybody's diet. She lives in the Netherlands.

VEGGIE IN PUMPS

Through a shared love for sustainability and plant-based cooking, our friendship grew into an ecofabulous lifestyle consultancy firm, Veggie in Pumps: enjoying the good life in style while respecting humans, animals, and the environment. It became our mission to widen the appeal of plant-based cuisine to a large group of consumers and, later, to put seaweed on the map. We learned that enhancing our food with seaweed could help mitigate several of the world's ecological problems, enable a more fair distribution of food, and reduce animal suffering, and that seaweed makes efficient use of natural resources while being beneficial for the human body. These reasons inspired us to make this enticing cookbook about algae and seaweed to give them a central role in our food's future.

Roos Rutjes & Lisette Kreischer

MARCEL SCHUTTELAAR, a nutritional engineer, is the founder of the North Sea Farm Foundation, the engine behind the cultivation of seaweed in the North Sea. He lives in the Netherlands.

NORTH SEA FARM FOUNDATION

Right now, we are at the tipping point of seaweed cultivation in the North Sea—the start of a promising new social and environmentally conscious economy. Seaweed is sustainable, healthy, and a shining example of Dutch agrifood and maritime expertise. Our goal when starting out was to harvest the first pound of offshore-cultivated seaweed in the Dutch North Sea by mid-2015. To achieve this, we established the North Sea Farm Foundation, a platform organization with its own offshore testing location, which we put at the disposal of researchers and entrepreneurs. We share results and experiences within our organization. Currently, we are working with our partners to develop the first commercial offshore seaweed farm. And when, a couple of years from now, we are looking out over the North Sea from the shore, we will be able to say: Out there, something extraordinary is growing. . . .

Marcel Schuttelaar, Job Schipper, Eef Brouwers, Koen van Swam & Marlies Draisma

---

This book has been made possible by Schuttelaar & Partners, a consultancy firm for a healthy and sustainable world. One of its initiatives is stimulating cultivation and consumption of seaweed; *Ocean Greens* **is part of this project. For more information: schuttelaar.nl.**